Big Science Secrets, Lies, and Mistakes

Shockingly Simple Truths the Public Has Never Been Shown

———————

Ted Huntington

Big Science Lies, Secrets, and Mistakes
Shockingly Simple Truths the Public Has Never Been Shown
Copyright © 2012 by Ted Huntington

ISBN (978-0-9881922-2-5)

www.tedhuntington.org

Dedicated to the many people tricked by lies

Contents

Introduction..7

Light is made of material particles and is the basis of all matter 8

The universe is infinite in size and age, the "big bang" and "expanding universe" theories are probably false 19

Globular clusters are made by advanced living objects and are the inevitable result of natural selection of the best adapted species- our future is to try to build a globular cluster 26

The big picture: nebula to spiral to globular ...28

Where are the bipedal (two-leg walking) artificial muscle robots that can shop, cook, and clean?...30

Any atoms can be fissioned and fusioned......31

Matter and motion are apparently completely separate and cannot be exchanged...............33

The electromagnetic theory of light is almost certainly false 34

"Remote neuron reading and writing" and camera micro and nano technology may have been developed to a very advanced state many years ago without the public being shown or even told about it 36

Diffraction, interference, polarization, refraction, and double refraction are all actually particle reflection...41

The neutron is probably a hydrogen atom 51

Why no telling the public about the details of evolution, the history of science, and the future? 52

"Non-Euclidean" surface geometry unlikely to apply to the universe 53

The special and general theories of relativity are very unlikely 56

So what is the most accurate interpretation of the universe in my view? 63

Timeline of Science 65

Other Important Ideas 94

First strike violence (FSV) is the big evil but somehow the public hasn't realized it 95

Teach the public the details of evolution, the history of science, of the future, and of religions 99

Full and constant democracy where people get to vote directly on the laws they must live under 103

Ending forced labor 108

Ending discrimination based on age 109

Total freedom of all information-no jail or fine for any info owned- the myth of "privacy" 110

Decriminalize recreational drugs 115

Decriminalize consensual adult prostitution . 119

Antipleasure ferver 123

Consent-only health care (ending torture and unconsensual experimentation in the psychiatric hospitals) 131

A "one-letter-equals-one-sound", one-stroke, easy-to-write, democratically-determined phonetic alphabet 138

The other species, their thoughts, vegetarian alternatives 141

An actual logical non-pseudo-science way to lose weight 142

Reality of bipedal robots doing all manual labor 144

Other popular mistaken theories and beliefs 145

Mistaken Belief: That God or Gods exist..... 146

Mistaken Belief: That a Heaven or Hell exists 148

Mistaken Belief: That a Devil exists 150

Mistaken Belief: That Jesus was the son of God, or a part of God, or was supernatural.. 151

Mistaken Belief: That Muhammad was a profit of God 153

Mistaken Belief: That many claims of Judaism, Buddhism, Hinduism, and other religions are accurate and useful 154

Mistaken Beliefs: Superstitions 156

Many other mistaken popular beliefs 157

Conclusions – What can we do?.................. 159

Introduction

There are some important science truths that I have learned that I think will help people tremendously in understanding the universe.

I have a feeling that there are many simple science truths and advanced technology purposely and deliberately not being shown to the public. In particular, there is the possibility that dust-sized cameras and other devices that float around inside houses and bodies have possibly been developed. In addition, there is the possibility that "remote neuron reading and writing" (RNRAW), which I realize sounds startling, and which I discuss in more detail later, has been developed long ago in the past, and has possibly resulted in a system of "direct-to-brain windows" (D2BW) which, although it sounds unusual, many people may be receiving while, sadly, the vast majority of the public may be denied and excluded from receiving and enjoying. The realization that such a secret segregation may exist is one reason why the excluded public has not received most of these simple science truths. Beyond that, greed and paranoia often serves to keep many scientific advances a secret for many years. There simply are many "big lies", and exposing them, collecting evidence against them, determining the truth, and showing it to the poor public takes a lot of time and effort.

In this book I am just going to summarize quickly these simple science truths. I am by no means an expert, and I do not claim to have all the answers, or to be 100% accurate all the time, but many of these truths are just shockingly simple to understand and verify.

Light is made of material particles and is the basis of all matter

It seems very likely that light is made of material particles, and is not a transverse electromagnetic wave as is currently claimed. All galaxies, stars and planets are corpuscular, so it seems logical that light is made of corpuscular bodies too. In fact, all matter is probably made out of light particles, and the explanation is simple. When we light a candle, light particles from the candle and oxygen in the air are emitted into our eyes from the flame, and as time continues, the candle is made smaller in size. The only logical conclusion is that the light particles were in the candle (and oxygen) the entire time, and that the candle and all atoms are simply made of light particles – that the light particle (the "photon") is the basic atom of all matter. This is true even for the so-called "antimatter" because when a proton and antiproton collide, the result is not empty space, but, instead, of course, all matter is conserved and emitted as light particles and other larger fragmentary particles also made of light particles. Even "anti" particles are made of photons. All other particles besides photons (including atoms and molecules) are "composite particles". This includes all the sub-atomic particles (mesons, etc.) too- all made of light particles. The claim of relativity that light is "massless" is absurd, in particular when we realize that light is the basis of all matter, the most fundamental atom that all matter is made of. Perhaps even light particles are made of smaller particles.

It's useful to just quickly summarize a little history about the various explanations of light given to the public over the last few centuries. Long ago, around 56 BC, the Roman writer Lucretius described light as being made of atoms that move very fast[1]. The collapse of science and the rise of the Dark Ages delayed progress until around 1228 AD when the first president of the newly established Oxford University published a book theorizing that all matter is made of light[2]. So this idea of light being made of material particles, and being the basis of all matter, is not only not new, but is really pretty ancient. A more precise explanation of light did not reach the public until 1664 when René Descartes' book "Le Monde" was published, 18 years after his death. In "Le Monde" Descartes revives the theory that light is made of particles, comparing light to a ball, and is the first to describe the two major theories of light: the wave and corpuscular theory[3]. So Descartes is really a hero for comparing light to a ball. It's interesting to note that the earliest "wave" theories are "corpuscular wave" theories; the wave is made of a material particle medium. This initial "wave" theory is probably more accurately called the "constant collision" theory, because in this theory light is transmitted by the constant collision of material particles that fill space, while the "corpuscular" theory is probably more accurately called the "rare collision" theory, because in this theory light is transmitted by particles that move through mostly empty space. It may be that inside

[1] Titus Carus Lucretius, "T. Lucreti Cari De rerum natura libri sex, Volume 1", 1866, lines 176-229, p530
http://books.google.com/books?id=oiUTAAAAQAAJ

[2] Robert Grosseteste, tr: Clare C. Riedl, "On Light {De Luce}", 1942.
http://web.mit.edu/jwk/www/docs/Riedel%201942%20Grosseteste%20On%20Light.pdf

[3] Descartes, R. Le Monde ... Ou Le Traité De La Lumière Et Des Autres Objets Principaux Des Sens, Avec Un Discours De L'action Des Corps Et Un Autre Des Fièvres, Composez Selon Les Principes Du Même Auteur. Michel Bobin et Nic. le Gras, 1664, Chapters 13 and 14.
http://books.google.com/books?id=DHEPAAAAQAAJ
English translation:
Rene Descartes, Translated by Michael S. Mahoney, "The World or Treatise on Light", Chapters 13 and 14.
http://www.princeton.edu/~hos/mike/texts/descartes/world/worldfr.htm

dense objects like planets and stars light particles constantly collide without moving much (like the wave theory describes), but when light particles reach the surface and empty space they rarely collide (like the corpuscular theory describes). In any case, three years later in 1667 Robert Hooke[4] firmly establishes the so-called wave theory for light, in which light is a *motion* through a medium of constantly colliding particles. Hooke adapts Descartes' theory for light, but introduces the theory that the medium for light is a liquid that exists between the stars and planets which he connects to the ancient theory of *aether (αἰθήρ)*.[5] Way back in the 300s BCE, Aristotle had added aether as a fifth element, in addition to earth, fire, air, and water (which eventually evolved into the more than 100 elements on the Periodic Table today). Initially, in Aristotle's view, aether was supposed to be what the Heavens (the realm of the stars) are made of.[6] In 1672[7] Isaac Newton publishes an alternative interpretation of Descartes' theory of light to more clearly and firmly establish the so-called "corpuscular" theory for light. In this view light is made of material particles (corpuscles) that *move*

---

[4] Hooke, R. Micrographia: Or, Some Physiological Descriptions of Minute Bodies Made by Magnifying Glasses. With Observations and Inquiries Thereupon. printed for James Allestry, 1667, p56-57.
http://books.google.com/books?id=SgFMAAAAcAAJ&pg=PA56

[5] Hooke, R. Micrographia: Or, Some Physiological Descriptions of Minute Bodies Made by Magnifying Glasses. With Observations and Inquiries Thereupon. printed for James Allestry, 1667, p96-97.
http://books.google.com/books?id=SgFMAAAAcAAJ&pg=PA96

[6] Aristotle, John Leofric Stocks, "On the Heavens" (Oxford: Clarendon Press, 1922), Book 1, Chapter 3.
http://ebooks.adelaide.edu.au/a/aristotle/heavens/index.html

[7] Isaac Newton, "A Letter of Mr. Isaac Newton … containing his New Theory about Light and Colors", Feb 19, 1671/2, in English, c. 5,263 words, 13pp. Published in: Philosophical Transactions of the Royal Society, No. 80 (19 Feb. 1671/2), pp. 3075-3087.
http://www.newtonproject.sussex.ac.uk/view/texts/normalized/NATP00006

through any medium. Newton compares the motion of light to a tennis ball, just as Descartes did. Many people recognize Isaac Newton as being the founder of the theory of gravitation, but few people know that Newton heroically established the corpuscular theory of light, which is perhaps of equal importance to the theory of gravitation, especially when we realize that material particles of light are probably the basis of all matter. So initially everybody agreed that light is made of material particles, the disagreement was only how light is transmitted: by many collisions of particles as a wave like sound in air, or by particles moving mostly through empty space like a ball in air. The corpuscular (or mostly empty space, rare collision) theory of light, in my view the more logical and accurate theory, held popularity for about 100 years. This period was, in my opinion, a bright and progressive period for science. But all that changed, at least in terms of physics. The rebirth of logical and intuitive science of the 1700s collapsed in the early 1800s, and we still live in the Dim Era that resulted. In 1801, Thomas Young was the first person to publish the frequencies (and wavelengths or particle intervals) of various colors of light[8], which is a great achievement, but Young supported a wave theory for light. In the same paper, Young advanced the idea that rays of light "interfere" with each other, an effect for light similar to that for sound where two sounds will cancel to produce places of silence- the analogy being that two waves of light may cancel to produce places of darkness (which I examine in more detail later). Young (in 1817[9]) changed the

[8] Thomas Young, "The Bakerian Lecture: On the Theory of Light and Colours", Philosophical Transactions of the Royal Society of London (1776-1886),Volume 92, (1802), pp12-48.
http://books.google.com/books?id=-XAXAQAAMAAJ&pg=PA140
[9] "Letter from Dr. Young to M. Arago", Jan. 12, 1817, found in: Young, T., G. Peacock, and J. Leitch. Miscellaneous Works: Scientific Memoirs. Murray, 1855.
http://books.google.com/books?id=-XAXAQAAMAAJ&pg=PA380

traditional wave theory to claim that light is not a forward and backward ("longitudinal" or "point") wave as Descartes and then Hooke had described, but instead is a "transverse" wave; a wave with an amplitude. Young felt that a transverse wave could better explain the phenomenon of light interference. Augustin Fresnel (in 1821[10]) also sided with this transverse wave theory. Sadly, this (in my opinion) unlikely wave or "undulatory" theory replaced the more logical corpuscular theory in popularity. As an aside, many people probably don't know that Young is credited with being the first, in 1807[11], to formally apply the word "energy" (the "currency" of scientific theories from the 1800s on) to Leibniz's "vis-viva" ("living force") quantity mv^2 of 1695[12]. Moving on with the story, James Clerk Maxwell, in 1864 (and 1873), accepted and adjusted Young's transverse wave theory by supposing that light is made of not one transverse sine wave in an aether, but instead, is made of two waves: one an electric wave, and the other a magnetic wave, both positioned at 90 degrees to each other.[13,14] And this (in my humble view) very unlikely "electromagnetic" interpretation

[10] A. Fresnel, 'Considerations mecaniques sur la polarisation de la lumiere', Oeuvres, Vol. I, 629-49; Annales de chimie et de physique, Vol. XVII (cahier de juin 1821), 167 ff, p168.
http://books.google.com/books?id=lrc-AAAAcAAJ&pg=PA629

[11] Young, T. A Course of Lectures on Natural Philosophy and the Mechanical Arts. Johnson, 1807. A Course of Lectures on Natural Philosophy and the Mechanical Arts., p78.
http://books.google.com/books?id=YPRZAAAAYAAJ&pg=PA78

[12] Gottfried Leibniz, "Specimen Dynamicum" (1695).
http://books.google.com/books?id=0je_DN18UkoC&pg=PA315
English translation: L. E. Loemker, "Philosophical Papers and Letters", (1976), pp.435-452.

[13] James Clerk Maxwell, "A Dynamical Theory of the Electromagnetic Field", Royal Society Transactions, Vol. 155, 1865, p. 459-512.
http://books.google.com/books?id=xVNFAAAAcAAJ&pg=PA459

[14] James Clerk Maxwell, "A treatise on electricity and magnetism.", 2 vol., 1st ed, Oxford, 1881, p383-398.
http://books.google.com/books?id=gmQSAAAAIAAJ&pg=PA383

has, shockingly, held to this sad day. For example, the spectrum of light is still called the "electromagnetic spectrum". But the corpuscular side fought back and in 1881 Albert Michelson, a "Master" in the US Navy, reported an experiment that showed that the speed of light is unchanged relative to the motion of the Earth around the Sun through the supposed aether medium. If there is a medium for light, and it is stationary relative to the motion of the rest of all the matter in the universe, then this motion should make the light waves take less time in the direction of the Earth's motion around the Sun, but this wasn't found. Michelson was even so bold as to write that "The result of the hypothesis of a stationary ether is thus shown to be incorrect, and the necessary conclusion follows that the hypothesis is erroneous.".[15] And then giving a final stamp of approval from none other than the D2BW specialist himself Alexander Graham Bell writing: "In conclusion, I take this opportunity to thank Mr. A. Graham Bell, who has provided the means for carrying out this work". Without any aether medium, it is hard to imagine light as a wave. This made the corpuscular theory for light appear to be the more accurate interpretation. But like the Empire, the wave camp struck back. To explain this experiment of 1881 (and the mysteriously more popularized experiment of 1887[16]) and to try to save the theory of light as a wave that moves through an aether medium that fills the universe, George FitzGerald, in 1889[17] developed a theory that matter

---

[15] Albert A. Michelson, "The relative motion of the Earth and the Luminiferous ether", The American Journal of Science, Volume 122, 1881, p120.
http://books.google.com/books?id=S_kQAAAAIAAJ&pg=PA120

[16] Albert A. Michelson and Edward W. Morley, "On the Relative Motion of the Earth and the Luminiferous Ether", American Journal of Science, s3, v34, Num 203, 11/1887, p333.
http://books.google.com/books?id=0_kQAAAAIAAJ&pg=PA333

[17] George FitzGerald, "The Ether and the Earth's Atmosphere.", Science, Vol 13, Num 328, 1889, p390.

contracts in the direction of motion just enough to account for the absence of any apparent difference in the speed of light that would be due to the movement of the Earth through a stationary aether. In 1892 (and 1899) Hendrik Lorentz threw his support behind this theory and expanded it by saying that not only does matter contract in the direction of motion just enough to offset it's motion through a stationary aether, but that even time itself can be contracted or dilated depending on the motion of matter[18,19]. Lorentz theorized that, instead of there being one time for all of the universe, each piece of matter has it's own time. Most of us can probably recognize these theories as being doubtful. Michelson rejected these unlikely explanations even as late as 1927 writing: "Lorentz and Fitzgerald have proposed a possible solution of the null effect of the Michelson-Morley experiment by assuming a contraction in the material of the support for the interferometer just sufficient to compensate for the theoretical difference in path. Such a hypothesis seems rather artificial, and it of course implies that such contractions are independent of the elastic properties of the material."[20]. The next big

http://books.google.com/books?id=8IQCAAAAYAAJ&pg=PA378

[18] H. A. Lorentz, "The Relative Motion of the earth and the Ether", Konink. Akademie van Wetenschappen te Amsterdam, Verslagen van der gewone Vergaderingen der Wis- en Natuurkundige Afdeeling, 1892, 1:74 ff; also in H. A. Lorentz, Collected Papers (The Hague: Martinus Nijhoff, 1937), vol 4., pp219-223.

http://books.google.com/books?id=8Q9WAAAAMAAJ

[19] H. A. Lorentz, "Théorie simplifiée des phenomènes electriques et optiques dans des corps en mouvement.", Traduit de Versl. K. Akad. Wetensch. Amsterdam 7, 507, 1899.

"Simplified Theory of Electrical and Optical Phenomena in Moving Systems", Proceedings of the Royal Netherlands Academy of Arts and Sciences, 1899 1: 427-442.

http://en.wikisource.org/wiki/Simplified_Theory_of_Electrical_and_Optical_Phenomena_in_Moving_Systems

[20] Albert Michelson, "Studies in Optics", Chicago University Press, 1927, p156.

development happened in 1900 when Max Planck partially revived the particle theory for light by theorizing that objects emit light in individual units he called "energieelements" but which were later called "quanta". Planck created a simple equation E=hv where E is an Energieelement, h is a constant and v is the frequency of emitted light[21]. In my view, Planck's "quantum" theory will be remembered mostly for partially reviving a particle theory for light, because most of the "energy" work of the 19th and 20th century, like the theories of relativity, are abstract and overvalued, especially when looking at the secret and more practical development of robots, rockets, transmutation, RNRAW, and nanoscale particle devices.   Albert Einstein was an early supporter of Planck's quantum theory, and applied it to describe the photoelectric effect famously in March of 1905[22]. So Einstein initially appeared to be a supporter of a corpuscular theory for light. I had said that it is hard to imagine light being a wave without an aether medium, but Albert Einstein could and did. In June of that same year Einstein

---

[21] M. Planck, "Zur Theorie des Gesetzes der Energieverteilung in Normalspektrum," Verhandlungen der Deutsches Physikalisches Gesellschaft 2 (1900), pp. 237-245.
http://books.google.com/books?id=zYYMAAAAYAAJ&pg=PA237
and "Uber das Gesetz der Energieverteilung im Normalspectrum", Annalen Der Physik, 4 (1901), p553-563.
http://books.google.com/books?id=j6AqAAAAYAAJ&pg=PA553
English translation:
Max Planck, Alexander Ogg, "On the Law of Distribution of Energy in the Normal Spectrum", 1903.
http://theochem.kuchem.kyoto-u.ac.jp/Ando/planck1901.pdf
[22] A. Einstein, "Über einen die Erzeugung und Verwandlung des Lichtes betreffenden heuristischen Gesichtspunkt", Annalen der Physik (ser. 4), 17, 132-148.
http://www.physik.uni-augsburg.de/annalen/history/einstein-papers/1905_17_132-148.pdf
English translation: "On a Heuristic Point of View Concerning the Production and Transformation of Light"
http://users.physik.fu-berlin.de/~kleinert/files/eins_lq.pdf

published his famous "special theory of relativity" [23] which adopted the far-fetched aether-saving theories of FitzGerald and Lorentz that matter and time contract depending on the speed of a material object relative to a stationary aether, but in an ironic twist, and perhaps as a compromise with the (by that time) nearly extinct corpuscular school, Einstein rejected an aether medium for light as being "superfluous", choosing instead to base motion relative to an observer. In this view the mass of an object increases the faster it moves (gaining mass from some unknown source), and time, as experienced by the object, slows down the faster it moves relative to (presumably) the rest of the matter in the universe. Einstein also supposes that the speed of light is always constant (which Crenshaw, et al[24] and Pound and Rebka[25] disprove in 1960 by showing that larger gravity speeds up light particles which decreases their frequency). In September of 1905 Einstein publishes his famous $E=mc^2$ equation (originally $m=L/c^2$)[26], which, to Einstein's credit at

[23] A. Einstein, "Elektrodynamik bewegter Körper", Annalen der Physik (ser. 4), 17, 1905, 891-921.
http://www.physik.uni-augsburg.de/annalen/history/einstein-papers/1905_17_891-921.pdf
"On the Electrodynamics of Moving Bodies"
http://users.physik.fu-berlin.de/~kleinert/files/eins_specrel.pdf
[24] T. E. Cranshaw, J. P. Schiffer, and A. B. Whitehead, "Measurement of the Gravitational Red Shift Using the Mössbauer Effect in Fe57", Phys. Rev. Lett. 4, 163-164, 165-166 (1960).
http://prl.aps.org/abstract/PRL/v4/i4/p163_1 and
http://prl.aps.org/abstract/PRL/v4/i4/p165_1
[25] R. V. Pound and G. A. Rebka, Jr., "Apparent Weight of Photons", Phys. Rev. Letters, 4 (1960) 337.
http://prl.aps.org/abstract/PRL/v4/i7/p337_1
[26] A. Einstein, "Ist die Trägheit eines Körpers von seinem Energieinhalt abhängig?", Annalen der Physik (ser. 4), 18, 639â€"641.
http://www.physik.uni-augsburg.de/annalen/history/einstein-papers/1905_18_639-641.pdf
English transation: "Does the Inertia of a Body Depend upon its Energy Content?"
http://users.physik.fu-berlin.de/~kleinert/files/e_mc2.pdf

least suggests the theory that all matter might be made of light, but, to me at least, implies that matter and motion can be converted into each other, and that mass has some inherent motion, both of which I doubt. Then 10 years later in 1913[27], as if matter and time contraction and dilation depending on motion is not unlikely enough, Einstein (with help from Marcel Grossman) expanded this theory and adopted the radical and unlikely surface topology (so called "non-Euclidean") geometry (which I describe below), to try to explain the movement of matter in the universe in his famous "general theory of relativity". It is, in my humble opinion, and with all due respect, very unlikely that a restricted surface geometry applies to the movements of matter through time in the universe. But as radical, extreme, and unlikely as applying a surface geometry to the universe may seem to me, this explanation is currently the most popular. In 1922, Louis De Broglie joined Planck's $E=hv$ with Einstein's $E=mc^2$ to solve for the mass of what he called "atomes de lumière" (atoms of light). The mass of the light quantum is now described as it's "rest mass" because of the doubtful (and no doubt deliberately dishonest) claim that motion changes a body's mass. De Broglie didn't give an exact number but simply supposed that light atoms have a "masse très faible" (very low mass). This raises the startling truth, although I haven't scoured the archives, that no actual mass of a light particle in terms of grams has ever been made public. In 1909, Jean Perrin described the mass of the electron, taken to be 1000

---

[27] A. Einstein, M. Grossmann, "Entwurf einer verallgemeinerten Relativitätstheorie und eine Theorie der Gravitation. I. Physikalischer Teil von A. Einstein II. Mathematischer Teil von M. Grossmann", Zeitschrift für Mathematik und Physik, 62, 1913, 225–244, 245–261.
English translation: "Outline of a Generalized Theory of Relativity and of a Theory of Gravitation. I. Physical Part by A. Einstein II. Mathematical Part by M. Grossmann". In A. Einstein, Edited: M. Klein, et al, "The Collected Papers of Albert Einstein: Vol 4, The Swiss years: writings, 1912-1914", 1995.

times less than Avogadro's number, as 0.805×10^{-27} grams, so perhaps the actual mass of light particles in terms of grams is 1000 times smaller around 1×10^{-30} grams. It also raises the point that, really, the light particle is perhaps a better "base" unit of mass than an atom is. Perhaps we should describe small masses in terms of "number of light particles". Finally, around 1927 Arthur Compton coined the name "photon" for the "quantum of light"[28] (note that the term "photon" is currently not defined as a material particle, as I am saying it should be, so perhaps it should be officially redefined, or a new word should be created like "luxon" or "lighton" for the ancient interpretation of light as made of material particles). So that brings us to the present time where chaos still reigns and truth is still viewed as disease.

---

[28] A. Compton, "X-rays as a branch of optics", 12/12/1927.
http://nobelprize.org/nobel_prizes/physics/laureates/1927/compton-lecture.pdf

The universe is infinite in size and age, the "big bang" and "expanding universe" theories are probably false

It seems very likely that the universe is not expanding, and that there was no "big bang", as hard as that is to accept or understand for many people. The best reason for this that I can give is simply that there must be many galaxies so far away that not one particle of light emitted from them can reach us in the tiny part of the universe that we occupy. Any so-called background "radiation" (why do they never say background "light particles"?) can only be light particles from light sources that are near enough to reach us. When we build a bigger telescope and then receive light from more distance sources, will we then say that the universe just got bigger and older? Given this simple truth, it seems likely to me that the universe must be infinite in size, age and scale. Either theory, "big bang" or "Infinite Universe" are difficult to comprehend.

I find the "infinite universe" view, far more interesting and likely. In this view the universe is infinite is size and age, with no beginning or end in space or time. The universe is probably infinite in scale too. Stars, galaxies or even galactic clusters might be only light particles at some larger scale, and likewise, light particles at our scale, might be stars, galaxies or galactic clusters on some smaller scale. At any scale, light particles just bounce around forming different structures. In this sense, nothing really changes when a body is born, lives or dies- the same exact light particles just continue on as usual in their unknown course.

Beyond this idea that there must be matter so far away that not one particle from it can reach us, is the historical data involved with the expanding universe "big bang" theory – which many people have never been shown (for example, how many of you have ever seen figure 1 before?). Of course, because of the big neuron lie and many decades of silence about remote neuron reading and writing, you can be sure that corruption has played a major role in what "official" science theories are told to the public. The nicest thing that can be said is that this claim of a "red shift" is simply a mistake, but the evidence for D2B and the obvious and simple nature of the Bragg equation for gratings (which I will explain later), imply that this claim of Doppler shift was a very corrupt and intentionally dishonest claim that the authors knew was false, and that is mass marketed to the public like so many other famous claims (Oswald killed JFK, Sirhan killed RFK, 9/11 was 19 hijackers, etc.) while any other theory, in particular, the truth, is mostly banned from the market.

In the 1800s it was thought that the changing position of spectral lines could be used to measure the Doppler shift of celestial objects - how the frequency of light from a light source changes because of the relative motion of the light source toward or away from the viewer. Slipher was the first to measure the supposed Doppler shift of other galaxies by comparing the positions of two common absorption lines attributed to calcium, publishing it (without any photos) in 1913[29]. To my knowledge, the first actual image of these shifted calcium absorption lines was not made public until Milton Humason published some in a famous photo in

---

[29] Slipher, V. M., "The radial velocity of the Andromeda Nebula", Lowell Observatory Bulletin, vol. 1, pp.56-57. http://adsabs.harvard.edu/full/1913LowOB...2...56S.

1936[30] (see figure 1). Looking at this photo more closely we see the visible spectrum of five different galaxies (although only in black and white). The two calcium absorption lines are apparent on the left (blue) end of the spectrum. Now notice how the overall width of the spectrum becomes smaller for each more distant light source. A result of this reduction in spectrum size is that the calcium lines appear to be closer to the center of the spectrum-just as part of the red end of the spectrum appears to shift in the blue direction (to the left). Are we to believe that part of this galaxy (the blue end) is moving rapidly away from us, but another part (the red end) is moving rapidly toward us?!

As obvious as this observation is, no major astronomers have publicly stated "the claimed red-shift of these 1936 spectra is not accurate because the spectra have different widths". It seems very likely to me that most of this shifting is due to the different sizes of the spectra. But in addition, some of this shifting may be the result of the simple Bragg equation (as I will explain later), because for each particular frequency of light reflected from a grating, there is a specific angle of incidence required. As a result, from simple trigonometry, that angle must occur at different positions along the grating for two light sources of different distance (see figure 8). So this shifting of the calcium lines and the spectrum in general is not a shifting of *frequency* but most likely a shifting of spectral line *position* only because the light sources are different in size and distance. The 1936 spectra have an obvious problem with *scale*-the size of each spectrum is not equal. When we stretch out the spectra to make them all equal in width, most of the apparent shift disappears.

---

[30] Humason, M. L., "The Apparent Radial Velocities of 100 Extra-Galactic Nebulae", Astrophysical Journal, vol. 83, p.10, Jan 1936.
http://articles.adsabs.harvard.edu//full/1936ApJ....83...10H/0000011.000
.html

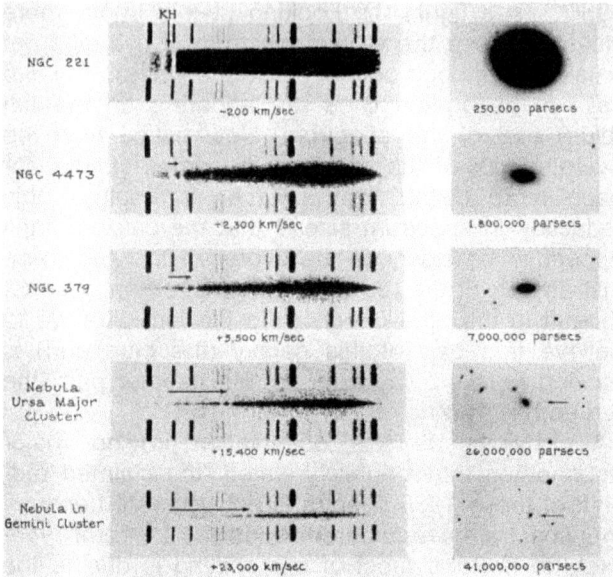

Figure 1. Image from a 1936 Milton Humason paper. Notice how the spectrum size gets smaller for each smaller light source, which in turn moves the two absorption lines closer to the center of the spectrum. In the bottom spectrum, is part of the galaxy coming rapidly at us because the right (red) end is blue-shifted?

Then twenty years later (after World War 2, in 1956), Humason, this time with astronomers Mayall and Sandage, mysteriously revisit the "red-shift" claim after 20 years of silence, and publish a second infamous photo (see figure 2). Unlike the first 1936 photo, not only are the calcium absorption lines in this photo much more difficult to see, but their supposed position relative to the rest of the spectrum is far from their appearance for the galaxies in the 1936 photo. The only other photos of the full visible spectrum with the calcium absorption lines that I have ever seen are derived (apparently) from these 1956 spectrographs. These are the famous 5 spectra that are printed in some (even modern) Astronomy textbooks and most likely the spectra shown in Carl Sagan's 1984 PBS television series "Cosmos" (see my paper on the web to see that image[31]). The "Cosmos" photo has apparently been "colorized" because there is no way that the full spectrum of a galaxy can have no color "red" in it. But if we then "re-colorize" it, making the right-most part red, then it is apparent that the frequencies of this spectrum do not match the

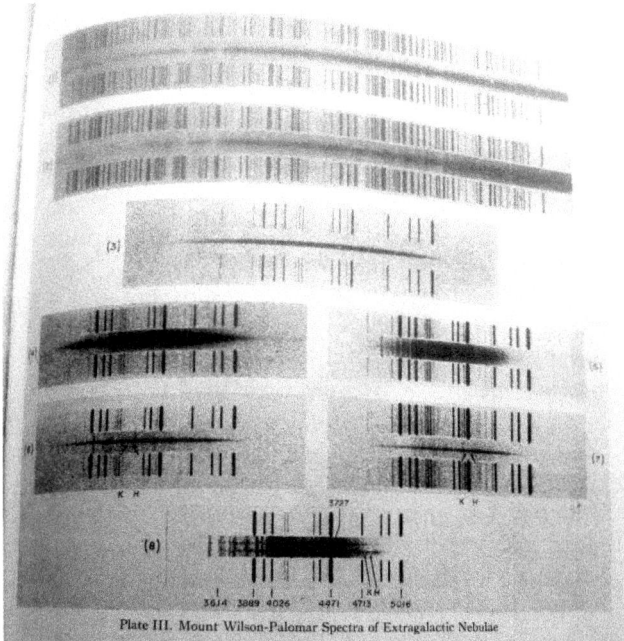

Plate III. Mount Wilson-Palomar Spectra of Extragalactic Nebulae

Figure 2. Image from Humason, Mayall and Sandage's 1956 paper, where a much bigger shift is claimed. I see the two absorption lines in the top two spectra in their usual position, but in the bottom three, a pair of similarly clear absorption lines are not apparent. If this was a pregnancy test, I would want another.

---

[31] Ted Huntington, "Spectral line position depends on distance of light source - Bragg Equation Effect", 05/10/2011.
http://tedhuntington.com/paper_Bragg.htm

frequencies of the reference spectrum (Hydrogen), for which red is farther away to the right. So here again, we are confronted with the truth that spectral lines change position depending on the size and distance of the light source, just as the Bragg equation requires.

If I am wrong: where are all the color photos like Humason's black and white 1936 (figure 1) and 1956 (figure 2) images showing the public the shifted calcium absorption lines? I have yet to see any other images of the supposed shifted calcium absorption lines other than those two from Humason in 1936 and 1956. Even a modern astronomy book like "Astronomy: a beginner's guide to the universe"[32] (2004) includes the same exact above spectra images published back in 1956- is there no more modern or even *any* other image of the shifted calcium absorption lines? This suspicious photo is becoming iconic and representative of fascism (facilitated by D2BW) in the United States, like the altered "Life" magazine cover photo of Lee Harvey Oswald[33], the "fall-guy" for the actual JFK murderer. Perhaps some amateur or even professional astronomers will be able to provide more spectra of those same galaxies Humason published either showing or not showing calcium absorption line shift.

You can see in figure 2 that not all galaxies have the iconic calcium absorption line pair in their spectra. How many people knew that? In fact, to the best of my knowledge, most galaxies do not have calcium absorption lines.

Note that whenever anybody doubts the expanding universe, insiders instantly present the only other alternative as being the "steady-state" universe theory, an unlikely theory (with all due respect)

---

[32] Chaisseon, McMillan, "Astronomy, A Beginner's Guide to the Universe", 2004, p422.
[33] "Life" (Feb. 21, 1964).

created by the astronomer Fred Hoyle in 1948[34], in which matter is constantly being created (from empty space) and destroyed in the universe. The simple answer to a "continuous creation" or "steady-state" theory is that there is no need to create or destroy matter when all matter is made of light particles that are never created or destroyed but simply move around in the space of the universe. Again, D2BW consumers all know this, but are paid and/or coerced by their D2BW dealer to "play dumb" and pretend to be completely unaware of a "matter was always here and is never created or destroyed" universe theory.

[34] Hoyle, F., "A New Model for the Expanding Universe", Monthly Notices of the Royal Astronomical Society, Vol. 108, p.372.
http://adsabs.harvard.edu/full/1948MNRAS.108..372H

Globular clusters are made by advanced living objects and are the inevitable result of natural selection of the best adapted species- our future is to try to build a globular cluster

Like the simple truth about light being a material particle, the immense value of remote neuron reading and writing, and bipedal robots (which I will talk about soon), I have to scratch my head and wonder why the popular people in science and education have remained silent for centuries, not even entertaining the public with the idea that globular clusters are made by living objects. Globular clusters (see figure 3) are groups of stars found around all spiral galaxies. Clearly, we are going to build cities on the moon of Earth, on Mars and the other moons and planets. Our population is going to continue to grow and grow in number if we are successful. And of course, we are going to send ships to and build cities around the other stars. Gravitation provides a very simple way to pull stars closer together- by simply using another mass to pull them along in any direction wanted (although the pulling object would have to be very massive, the better method is probably using many coordinated smaller masses). Given this truth, when we go to those other stars and pull them closer, to save fuel and time in moving back and forth between them, we would look a lot like a globular cluster. Just like bacteria and humans, probably living objects around globular clusters do exactly the same thing, convert matter into more of them. Like us, living objects around the stars in globular clusters probably have a strong desire to explore the universe.

If we do not drop our mistaken beliefs and start to collect other stars soon, then our star will eventually be "pulled in" to be consumed by some other nearby cluster like "Hyades", which already has around 300 stars, and is located only 150 light years away.[35]

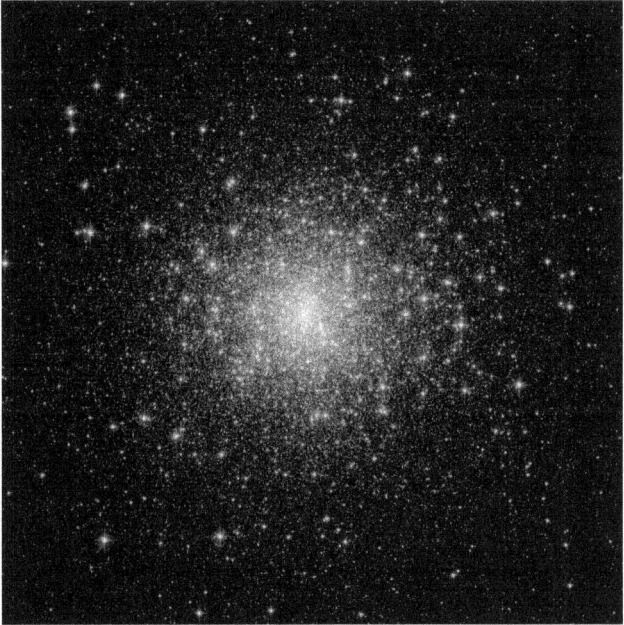

Figure 3. Globular cluster M15 in the Milky Way Galaxy: one end product in the survival of the fittest. Our future is to build a globular cluster.

[35] Perryman, M.A.C., et al. (1998). "The Hyades: distance, structure, dynamics, and age". Astronomy & Astrophysics 331: 81–120. http://arxiv.org/pdf/astro-ph/9707253v1.pdf

The big picture: nebula to spiral to globular

An interesting related note to this truth about globular clusters is that there is a simple pattern in the universe (see figure 4): light particles move to empty spaces, become trapped with each other, growing to form atoms. Eventually as more light particles become trapped, a nebula is formed, then a spiral galaxy. Living objects evolve around stars in the spiral galaxy and start to convert matter into more of them, pulling stars close together as they colonize them. Eventually the galaxy is all globular and is what is being called an "elliptical galaxy" (although I think "globular galaxy" is probably a better term). It may be that a globular galaxy can exist for a long time by consuming other galaxies or existing on incoming light particles. Natural selection operates everywhere in the universe, and galaxies and globular clusters may be viewed like fish in an ocean of space- the most adapted survive, the less adapted disintegrate. If most globular galaxies emit more light than they take in, then they would ultimately emit the light particles stored in stars and be constantly reduced in size until they were almost empty space again.

This basic pattern seems logical to me, but no major astronomers or other scientists are saying this. They tell the public that globular clusters are made of "generation two" stars. They say nothing about a globular cluster being a massive matter conversion center made by very adapted living objects where any and all available matter is pulled in, consumed, and changed into more of them. But if they are D2B consumers, how could they not know such a simple truth?

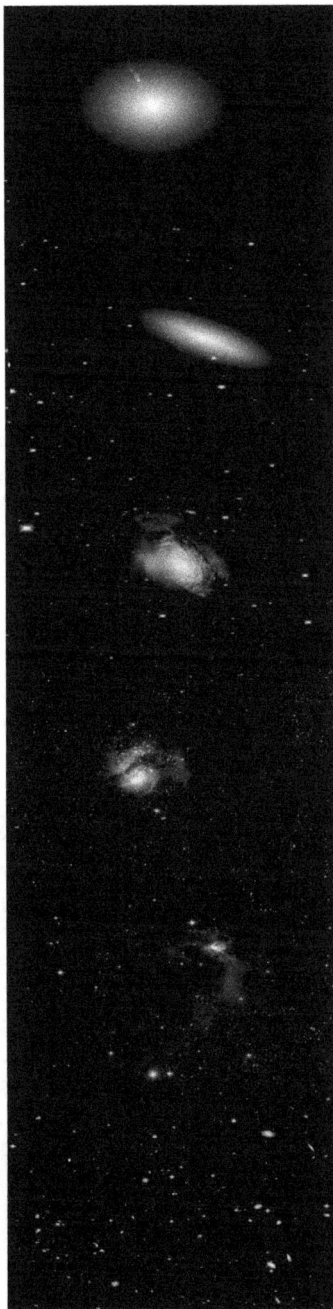

Figure 4. Stages in the evolution of galaxies (from left to right): Empty space, nebula (LDN1622), nebula (M8), (NGC3521), "Lenticular"- between spiral and globular (NGC3115), globular (M87).

Where are the bipedal (two-leg walking) artificial muscle robots that can shop, cook, and clean?

First the bipedal robots. Are we to believe that people who mass produce personal computers could not build walking robots by now? Come now. Knowing that many people probably figured out how to see, hear and send images and sounds to and from brains long ago, adds to the doubt. Beyond that, it seems unusual that the simple contraction of muscle, which the majority of animals do all the time, has not been achieved in an "artificial muscle" by the many thousands of chemists in multibillion dollar governments, universities and companies just like synthetic rubber and fabric was. The much more likely truth is that these inventions (very smart and fast moving bipedal robots and even light-weight artificial muscle robots) were probably developed many years ago, but those who control technology on Earth have somehow, out of unjustified fear, and perhaps greed, chosen to not allow these devices to enter the public. Think of how many lives could be saved by robot vehicle drivers with millisecond reaction times as opposed to the many mistake making humans that drive. Think of how much easier life would be for the average middle income person to have a robot to cook and clean for them. Perhaps part of the problem is the fear many old-world people have that robots will somehow kill off humans. This classic scenario is impossible in my mind, because, like particle devices, there is always a kind of stale-mate between two opposing sides. I think that robots will working for humans for a long time into the future– cooking, cleaning, etc., and will be the first to reach the other stars from Earth.

Any atoms can be fissioned and fusioned

Any atoms can be split (fissioned) and fused together (fusioned). The public isn't being told this, and so most people don't know this. I didn't know this until I found two papers from 1950.[36],[37] This process of changing one atom into another is traditionally called "transmutation". Apparently atoms can be broken down and built up using particle accelerators. In the first paper the authors explain that even atoms in the middle of the Periodic Table can be split apart. In the second paper, the authors explain how they build up larger mass atoms by colliding material with carbon ions. The traditional public story is that the quantity of atoms changed is so small for the quantity of electricity used, that the process is mostly useless. But it is obvious to me that transmutation is extremely important. Most moons and planets in this star system have very little free oxygen and hydrogen. The idea of transporting oxygen and water from Earth to other planets is very unlikely. Since all atoms are made of light particles, the goal is to make use of the tons of atoms right there on those distant moons and planets by converting the common atoms (like Silicon and Iron) into more useful atoms we need like Hydrogen and Oxygen. So just like the simple idea of all matter being made of light, of the value of remote neuron writing and artificial muscles, so it is with transmutation – just nearly absolute silence. The silence is one of the biggest indications of secrecy. Such ripe and valuable veins of scientific research

---

[36] Roger E. Batzel and Glenn T. Seaborg, "Fission of Medium Weight Elements", Phys. Rev. 79, 528 (1950).
http://prola.aps.org/abstract/PR/v79/i3/p528_1

[37] J. F. Miller, J. G. Hamilton, T. M. Putnam, H. R. Haymond, and G. B. Rossi, "Acceleration of Stripped C12 and C13 Nuclei in the Cyclotron", Phys. Rev. 80, 486 (1950).
http://prola.aps.org/abstract/PR/v80/i3/p486_1

would normally be massively explored and openly discussed as an important future goal. And most likely they have, but with a strict and most likely very harshly punished code of silence. Probably by now, just like remote neuron reading and writing nanotechnology, very efficient conversion of iron and other common atoms into a wide variety of more useful atoms has probably been achieved – but only for a group of perhaps a few million people that are in on the secrets.

Matter and motion are apparently completely separate and cannot be exchanged

The traditional view is that "energy" is the base form of all matter, and both matter and motion can be converted into each other. This is implied by any equation of energy ($E=1/2mv^2$, or $E=mc^2$, etc.). But simple logic shows that matter is clearly conserved, and that motion is clearly conserved, but matter cannot be converted into motion, and motion cannot be converted into matter. Conservation of matter is widely accepted as true by people, but not the principle of **conservation of motion** for some reason. The same is true for momentum ($p=mv$). Motion can be transferred from one piece of matter to another, but it seems unlikely to me that new matter or motion can ever be created or destroyed.

One reason for the silence about matter and motion being separately conserved may be because the phenomenon of acceleration is difficult to understand in terms of conservation of motion.

In 1876, James Clerk Maxwell published a book entitled "Matter and Motion"[38] which hints that this simple truth about matter and motion being separate was probably already known privately by 1876.

There is a comforting view that this possible truth provides: when a person you love dies, they are always still here, no part of them has disappeared from the universe; the light particles that they were made of simply move on to continue their mysterious and unknowable journey in the universe.

---

[38] James Clerk Maxwell, "Matter and Motion.", 1876.
http://books.google.com/books?id=6WgSAAAAIAAJ

The electromagnetic theory of light is almost certainly false

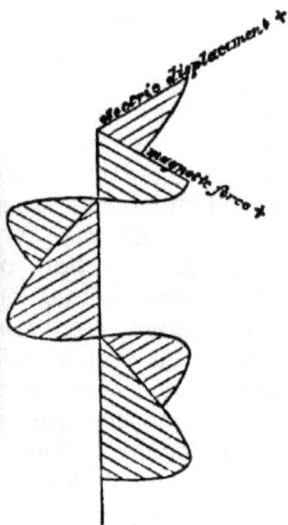

Fig. 66.

Figure 5. James Clerk Maxwell and a drawing of an "electromagnetic" light wave from his book of 1881[39].

In 1864, James Clerk Maxwell published his "electromagnetic theory of light" (fig. 5)[40], which is, with all due respect, a very unlikely theory, where light is supposed to be made of two sine waves: an electric wave, and a magnetic wave, which are at 90 degrees to each other, in an aether medium. Beyond

---

[39] James Clerk Maxwell, "A treatise on electricity and magnetism.",vol. 2, 1st ed, Oxford, 1881, p390.
http://books.google.com/books?id=gmQSAAAAIAAJ&pg=PA390
[40] James Clerk Maxwell, "A Dynamical Theory of the Electromagnetic Field", Royal Society Transactions, Vol. 155, 1865, p. 459-512.
http://books.google.com/books?id=xVNFAAAAcAAJ&pg=PA459

the Michelson experiment, in 1881, which is evidence against an aether medium for light, this theory is easily disproven by the fact that light with low frequencies, like radio, can be focused by a concave object (for example a mirror) to a point. It seems doubtful that light in the radio with a 10 meter or more wave length could be focused to a point of variable distance unless radio light has no amplitude. Beyond that, a straight line, material, particle interpretation for light is far less complicated and still can explain all known phenomena of light. The explanation of radio as light particles emitted from particle collisions in electric current seems an obvious but mysteriously missing alternative theory. Light is emitted from rubbing two objects together and from exothermic chemical reactions, but light is not called an "electromagnetic", "mechanical", and "chemical" wave. Light is better described as being made of material particles, and the so-called "electromagnetic spectrum" as the "spectrum of light particle frequencies".

Shockingly, this electromagnetic theory of light is still the most popular and "official" view nearly 150 years later. This work begins the "trick the public with math" years (and also the "integration and differentiation are the only tools to describe the universe" years). But this at least reawakens the idea that the public should gain access to invisible frequency particle communication (radio).

"Remote neuron reading and writing" and camera micro and nano technology may have been developed to a very advanced state many years ago without the public being shown or even told about it

Neurons are the cells that all nerves are composed of, and the vast majority of animals have nerves made of many neurons. Neuron reading and writing, that is, determining what the electrical value of a neuron is, or making a neuron cell fire, has a long history that most people have never been told about. For example, it seems unusual that the public has never even heard the phrase "neuron reading and writing", a phrase that, in my mind, seems to clearly and concisely describe this ancient science. This field mixes a variety of sciences including biology, electrical and mechanical engineering, and health science.

The many important details of neuron reading and writing, and its history are too massive to fit into a single chapter, so I have dedicated an entire book to the subject titled "Direct to Brain Windows: The History of Neuron Reading and Writing". But just to touch on some important points: many people are not aware that neuron writing has a long history that goes back at least to 1689 with Jan Swammerdam making a frog leg contract using two different pieces of metals.[41]

---

[41] John Joseph Fahie, "A History of Electric Telegraphy, to the Year 1837", E. & F. N. Spon, 1884.

In addition to direct neuron reading and writing, where electrodes are placed in physical contact to a body, there is the important field of "remote" neuron reading and writing, where neurons are read from and written to (or activated; made to fire) using light particles (photons). Remotely writing to neurons with invisible frequencies of particles goes back to at least 1791 and Luigi Galvani, who made a frog leg muscle contract using a remotely generated electric spark.[42]

Reading from neurons was publicly demonstrated by Richard Caton in 1875, who measured the electric currents in the brains of rabbits and monkeys during chewing, and in response to light shown in the eyes.[43] Remotely reading neurons was recently published by a team of scientists in Japan who captured and published images of what the human brain sees using only functional magnetic resonance imaging.[44]

Putting microscopic electronic devices inside individual cells that can be remotely communicated with to get information from or send information to a cell (for example to read the electric potential of a neuron or to make a neuron cell fire), is not a physical impossibility. Many people are not aware that in 2009 scientists in Spain succeeded in making an intracellular chip by putting nanometer scale silicon chips into living cells.[45]

---

http://books.google.com/books?id=0Mo3AAAAMAAJ

[42] Luigi Galvani, Elizabeth Licht, Robert Green, "Commentary on the Effect of Electricity on Muscular Motion", Waverly Press, 1953.

[43] Richard Caton, "The Electric Currents of the Brain", British Medical Journal, 1875, V2, p278.

[44] Miyawaki, Y., Uchida, H., Yamashita, O., Sato, M., Morito, Y., Tanabe, H. C., Sadato, N., Kamitani, Y. (2008). "Visual image reconstruction from human brain activity using a combination of multi-scale local image decoders. ", Neuron, 60, 5, 915-929.
http://www.cell.com/neuron/abstract/S0896-6273(08)00958-6

[45] Gómez-Martínez, Rodrigo et al. "Intracellular Silicon Chips in Living Cells." Small 6.4 (2010): 499–502.
http://onlinelibrary.wiley.com/doi/10.1002/smll.200901041/abstract

The question that should arise in many people's mind is: Is it possible that people have developed and deployed advanced technology much earlier, but failed to tell the public about it? There are classic examples, like the Manhattan Project, where a nuclear chain reaction was sustained, and explosive nuclear fission bomb developed without the public being told for years. It's clear that this technology, in particular remote neuron reading and writing, would give a group of humans an extraordinary military advantage over another group. If you can see and hear thoughts, and even send sounds and images to a brain remotely using invisible particles of light that nobody can see, all your opponent's plans could be seen and heard in their thoughts, and disastrous plans could be remotely sent there without your opponent knowing. In fact, it cannot be denied that even a massive secret "segregation" could have arisen, in which, one group of "lucky" people receives videos of other people's thought-images and thought-sounds, and videos of their intimate lives inside their homes, sent directly, harmlessly, and conveniently right to their eyes and ears, while a second group is left in complete ignorance, without even an idea that such technology is in widespread existence. You can see that this kind of secret invisible segregation could have a lasting selective effect on who gets the job, to reproduce, etc. and who doesn't.

So is there any evidence that, like the Manhattan Project, remote neuron reading and writing was developed and deployed secretly many years ago, and may be in wide-scale, but secret, use and abuse? I realize that this seems like a startling and doubtful claim, but believe it or not, *yes*, there is a lot of historical evidence to suggest that this is true.

Many decent people of the past have recorded many apparent hints that remote neuron reading and writing and microscopic camera technology may be a part of their lives. I detail most of these hints in my "Direct to Brain Windows" book, but I will give some here. The earliest potential hint I have found comes from a 1589 William Byrd poem[46] in which he writes "your minde is light". You can see that the word "light" is obviously meant to mean light in terms of mood as in a measure of weight, but there is a second possible meaning that "your mind is made of light particles which are emitted and captured as images and sounds". James Knowles suggested that brains emit signals as early as 1869.[47] Hugo Gernsbach was the first person I know of to suggest that recoding the sound of thought might be possible, all the way back in 1911[48], long before most of us were even born. A very explicit hint about microscopic intracellular devices and direct to and from brain communication can be found in the 1967 movie "The President's Analyst",[49] where a phone company robot gives a presentation where a tiny microchip enters a body and allows thought-calls directly to and from the brain.

Much of the evidence that micro and nanometer sized cameras and remote neuron reading and writing particle communication devices were secretly developed long ago is the silence. The absence of any public information about what seems to be a simple and important field of science is unusual. Why have we never heard of wireless nanometer sized cameras (nanocams) that float and fly around inside houses, or even of anybody talking about such things? What about tiny wireless intracellular devices? Are we to believe that nobody sees the

---

[46] W. Byrd, "Songs of Sundrie Natures", 1589.

[47] "Brain Waves: A theory", The Spectator, 01/30/1869.
http://ulsfmovie.org/docs_pd/Knowles_James_18690130.pdf

[48] Hugo Gernsback, "Ralph 124C 41 +", "Modern Electrics", Modern Electrics Publication, New York, Vol. 4, No. 3, June 1911. Taken from "Modern Electrics", Volume 3-4, Jan-Dec 1911, p165-168.

[49] Theodore J. Flicker , "The President's Analyst", Paramount Pictures, 1967

value in remote neuron reading and writing? It's hard to believe that nobody wants to talk about trying to see, hear, and send images and sounds to and from brains and to remotely contract muscles. In addition, it is obvious that receiving videos directly to our eyes and ears, removing the need for having to hold a screen, telephone, or headphone, with an equally harmless particle technology, is certainly the inevitable future. Why the silence?

 Beyond the explicit hinting, there is evidence from your own life. Probably the best daily evidence that remote neuron reading and writing may have secretly become highly advanced, and in wide spread use and abuse, may be demonstrated by "remote molestation", for example, when you hear a jeer in your ear, when one of your muscles involuntarily contracts (for example when you say something incorrectly), or when you thought you saw something for a second, but then looked again and it was not there.

 Like so many injustices of the past, trying to determine the many secret events surrounding the rise and development of neuron reading and writing, and camera technology, and making that information and all those videos public will probably take many centuries. Sadly, this realization and effort really has not even started in our time.

Diffraction, interference, polarization, refraction, and double refraction are all actually particle reflection.

Diffraction

Francesco Grimaldo (or Grimaldi) invented the term "diffraction" to describe the way that when light is passed through two holes, some light appears to "bend" and be seen outside the cone of light (see figure 6). This work was published in 1665.[50] The problem with this explanation is that Grimaldo never takes into account reflection of light from the side of the hole which definitely explains how light can appear outside of the cone of light.

One startling and suspicious truth is that Grimaldo's work, in which he is the first to define the new property of light called "diffraction", has, as far as I can see, never been translated into the English language. But yet, this ancient and important supposed property of light is widely accepted as fact. The "light bends around the corner" explanation has been accepted for centuries without even a public examination of the original claim.

The claim of diffraction is supposed to be evidence of a wave theory for light, and was thought to explain how light is spread out into the familiar color spectrum when colliding with a grating (a flat plate of glass with tiny equally-spaced lines carved into it). This theory resulted in the grating being called a "diffraction grating".

But in 1912 William Lawrence Bragg showed that so-called "diffraction" patterns can be explained as a

---

[50] P. Francesco Maria Grimaldo, "Physico-mathesis de lumine, coloribus, et iride", 1665, p1-11.
http://books.google.com/books?id=sZE_AAAAcAAJ

particle reflection phenomenon (see figure 7)[51]. However, apparently many people have (no doubt purposely) failed to show that the Bragg explanation and equation can also explain any so-called diffraction phenomenon. In Bragg's interpretation, particles reflect off the regular rows of atoms, just like the regular grooves of a grating, at specific frequencies defined by the angle of incidence and the space between atoms (or grooves).

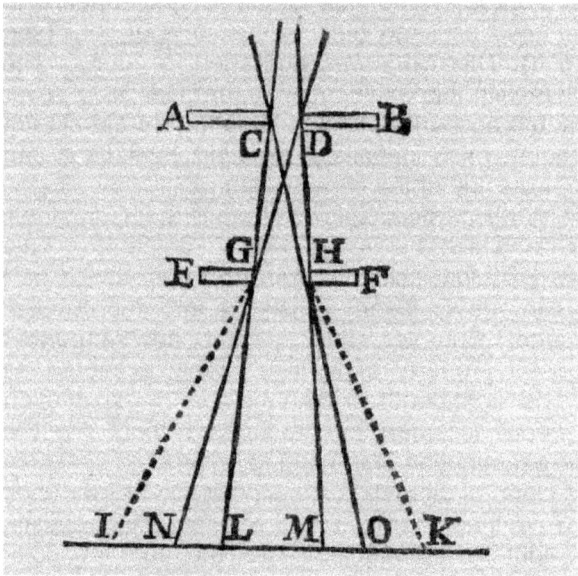

Figure 6. Image from Francesco Grimaldo's book "Physico-mathesis" showing what Grimaldo claims is a new property of light he calls "diffraction" But IN and OK are probably the result of particle reflection off the inside surface- think of a ray coming from above, reflecting at H and landing at I- lines of reflection are not drawn but must exist.

---

[51] Bragg, W.L. "The Diffraction of Short Electromagnetic Waves by a Crystal.", Proceedings of the Cambridge Philosophical Society, 1913: 17, pp. 43-57.
http://ulsfmovie.org/docs_pd/Bragg_William_Lawrence_19121111.pdf

Bragg made use of a simple equation $n\lambda=2D\sin\theta$ that he attributes to Schuster, but which is now called the "Bragg equation", to describe this property of gratings. In this equation n is the order of the resulting spectrum, λ the resulting reflected "wavelength" (or equivalently "particle spacial interval" or "distance between particles"), D the space between atoms (or grating grooves), and θ the angle the light makes with the atomic plane (or grating groove). So to create a particular frequency of light particles in the reflected spectrum, a precise angle and grating groove spacing is required. For example, green light (570 nm) on a grating with 1um between grooves requires an angle of incidence (with the groove) of 16°, but to reflect blue light (380 nm), a smaller angle of incidence, 11°, is required. So you can see that each different frequency of light in the colorful spectrum from a single light source comes from a different part of the grating.

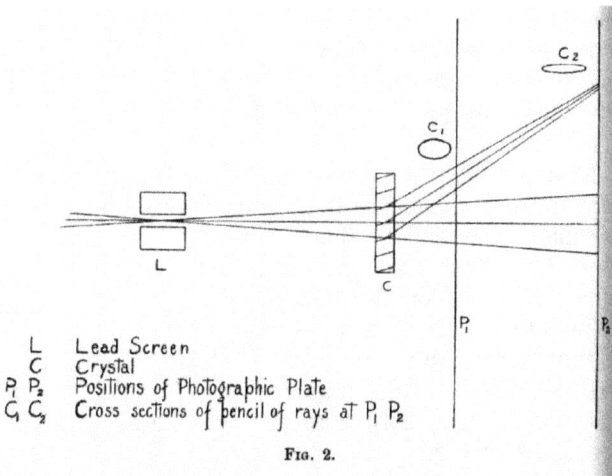

L	Lead Screen
C	Crystal
P_1 P_2	Positions of Photographic Plate
C_1 C_2	Cross sections of pencil of rays at P_1 P_2

Fig. 2.

Figure 7. Image from William Lawrence Bragg's 1912 paper "The Diffraction of Short Electromagnetic Waves by a Crystal" showing how diffraction can be interpreted as a particle reflection.

Because of this Bragg equation, simple trigonometry shows (see figure 8) that if two light sources are at different distances, the position on the grating of any particular frequency of light must be located at different places relative to the center.

A simple demonstration of this effect is to look through a $10 sheet of plastic grating you can buy on the Internet while walking toward a light. You can see that relative to the center, as you get closer to the light, the spectral lines move closer to the center (are "blue-shifted"), as you move farther away from the light the spectral lines move farther from the center (are "red shifted"). Another example is looking through one of those low-cost plastic "black tube" diffraction gratings at the Sun (only for a second so you don't hurt your eye). In order to see the colorful image of the Sun (not just the spectrum from ambient light- but the actual circular image of the bright Sun) in the black tube you need to look "next to the Sun" because when looking through the tube directly at the Sun, none of the rays of sunlight directly from the Sun can "make the angle" necessary (with the plastic grating near the eye) to reflect the frequencies in the visible spectrum.

The details of the particle reflection explanation for "diffraction" that are described by the Bragg equation are more complicated than you might expect for so simple an equation. Many different frequencies of light reflect off each grating groove surface, in very small quantity for each groove. But for many angles of incidence, there is a resonance between multiple beams over a group of grooves which reflect specific frequencies of light particles in a single specific direction. Many light sources, like stars, emit light particles in many directions from a central sphere, and this creates a progressively increasing angle of incidence from the center of the grating to the outer edge. All the frequencies of light that are reflected

Bragg equation: $n\lambda = 2D\sin(\theta)$
θ **is the angle of incidence required for any particular color**

For a grating with D=1000 lines/mm:
$\theta_{red\,(^-60\,nm)} = 22°$ **but** $\theta_{blue(360\,nm)} = 11°$

γ **= Groove angle to perpendicular**
$\alpha = 90 - (\theta + \gamma)$

$X_{s1} = Y_{s2}/\tan(\alpha)$
$X_{s2} = Y_{s1}/\tan(\alpha)$
$X_{s1}Y_{s2} = X_{s2}Y_{s1}$

The location of the spectral line changes depending on the distance to the light source.

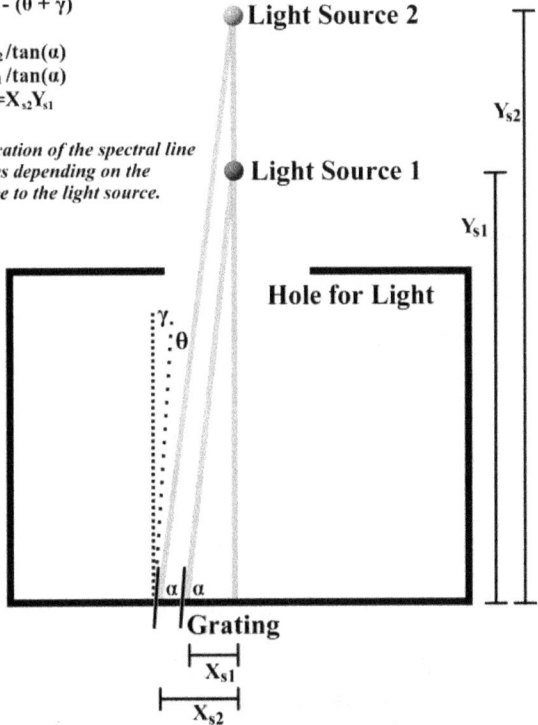

Figure 8. Simple trigonometry shows that two light sources at different distances cannot achieve the same angle with the grating grooves at the same location on a diffraction grating. This may explain some of the changing position of spectral lines from light sources of different distance. This effect is shown for two identical lamps at different distances in an image on the back cover of this book.

off a grating groove wall, for some specific angle of incidence, reflect in the same direction, but there is a stronger intensity (more light particles) over a few successive grooves, for one particular interval (wavelength) of light in one particular direction which is related to the angle of incidence and space between grooves. Other rays of light in between these resonant rays pass through without reflecting in the space between each groove. You can imagine the horizontal space between gratings as being rotated 90 degrees to become an identical vertical interval between light particles. Particles with a space between them that is related to the space between grating grooves, reflect off of a small group of successive grooves at the same time, in the same direction, to produce a signal strong enough to see. Not many light rays reflect this way, because the space where the angle of incidence and groove spacing align for any frequency is very small, so the spectrum of light is produced by a very small fraction of rays from the source light.

In a prism the color spectrum may also be the result of the angle "refracted" light makes with multiples of the regularly spaced planes of atoms in glass.

Bragg showed this equation to be true for x-rays, but it is obvious that this equation applies to all particles of matter (protons, atoms, molecules, etc.). This equation is true not only just for x-ray frequencies, but for all frequencies of particles (even largely spaced gratings for radio frequencies), and for any size particles- even sand grains and larger objects must produce the same phenomenon.

As an interesting footnote, I recently found that the University of London professor Herbert Dingle had recognized this change in spectral line position with light source distance based on the simple Bragg or "grating" equation over 50 years ago in a 1960 work.[52]

Interference

In his famous 1801 paper[53], Thomas Young formally introduced the claim that light waves interfere in a way similar to sound waves; where the waves can add or subtract from each other. The most traditional experiment used to demonstrate light interference is the "double-slit" experiment: a beam of light is sent through a double-slit which, when done correctly, will produce numerous lines, the dark areas being where light waves are supposed to "cancel out". But similar to Grimaldo's diffraction claim, reflection of particles of light off the inner sides of the slit (and off each other) are never accounted for. What many people aren't shown is that, a single-slit can produce similar lines, which implies that this distribution of light is a reflection phenomenon. In addition, we don't see light interference from two identical visible light sources (like two lamps in a room) without any reflecting slit apparatus. A simple experiment may show that light interference of two radio sources is only *additive*, not *subtractive*; in other words, unlike sound, with light interference, two signals can only add together, not subtract from each other. I only know of two publications that claim light interference for radio, one is an 1888 paper[54] of Heinrich Hertz, and the

[52] Herbert Dingle, "Relativity and Electromagnetism: An Epistemological Appraisal", Philosophy of Science, 27, (1960), p233-253.
http://www.jstor.org/stable/185967
[53] Thomas Young, "The Bakerian Lecture: On the Theory of Light and Colours", Philosophical Transactions of the Royal Society of London (1776-1886),Volume 92, (1802), pp12-48.
http://books.google.com/books?id=-XAXAQAAMAAJ&pg=PA140
[54] H. Hertz, "Ueber die Ausbreitungsgeschwindigkeit der electrodynamischen Wirkungen", Annalen der Physik, Volume 270 Issue 7, p551-569.
http://books.google.com/books?id=_D0bAAAAYAAJ&pg=PA551
English translation:

other is a 1897 paper[55] by Augusto Righi, for which there is no English translation that I am aware of. Hertz's claim is hard to visualize without video. Hertz finds that at certain points the signal from two sparks cause the two balls on a secondary conductor to have the same electric potential and so they do not spark, while everywhere else, they have a different potential and so do spark. But this fits with a corpuscular "additive-only" theory for light, since the light particles are still there, but simply in the same quantity in each ball. It's not an absence of signal; it's an absence of spark. Righi claims, in his work, that waves of radio light interfere just like visible light in equivalent optical experiments using thin films and mirrors, which to me implies a reflection phenomenon; a more convincing experiment would not require any reflecting apparatus, but would simply show how two light sources create subtractive interference just like two sound sources do. That Righi's work has not yet, as far as I know, been translated to English, and that no "radio interference" experiments have ever been reported by English speaking authors, shows a clear lack of effort to prove this point to the public.

Polarization
  Polarization is a similar reflecting phenomenon. Light entering polarizing material has its *direction* restricted (not, as is currently claimed by many authoritative sources, that rays of light all "vibrate" in the same direction). Polarizing material can easily be thought of as being made of long columns of planes

Heinrich Hertz, tr: D. E. Jones, "On the Finite Velocity of Electromagnetic Actions", "Electric Waves", 1893, 1962, p59.
http://books.google.com/books?id=EJdAAAAAIAAJ&pg=PA59
[55] Augusto Righi, "L'ottica delle oscillazioni elettriche", Bologna, 1897.
http://books.google.com/books?id=QRU6AQAAIAAJ
English translation: "The Optics of electrical oscillations".

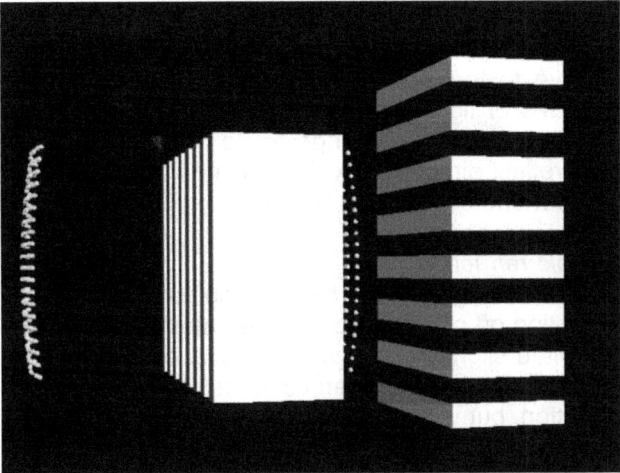

Figure 9. Image from a simulation showing light particles filtered by a vertical and then horizontal polarizer.

(see figure 9). When the polarizer is held vertically (the columns are vertical), any light rays with some sideways direction (usually referred to as the "X" dimension) are reflected and/or absorbed by the polarizing material, but the entire 180 degrees of vertical rays (with a "Y" dimension component) pass through unreflected. The opposite is true for a polarizer held horizontally; all rays with vertical directions are reflected back or absorbed by the polarizer walls, while rays in any of the 180 degrees of horizontal direction pass through. By putting two polarizers together at 90 degrees to each other, almost all ray directions are filtered out by reflection and absorption, except those passing directly through with no X or Y component to their direction (only a "Z" or forward component).

Refraction
Refraction is probably the result of light particles that are reflected by other light particles (within atoms) in a denser material. When this happens, the direction of the light particles changes; if Z is the

direction perpendicular to the refracting material, the X and Y components of the rays are made less relative to the Z. This is because of collisions the particles of light experience. Rays perpendicular to the material do not change angle because they can only reflect directly back (there is no X or Y direction component to lose from collisions).

Double refraction

Double refraction may be the result of light particles reflecting off of two perpendicular atomic planes that are tilted slightly in the X and Y planes so that light reflects off them and then continues through in the Z direction, but with a different X and Y angle. One set of rays is polarized relative to the Y and the other relative to the X direction, so two double refracting crystals can act like a polarizer in blocking most of the light with X and Y components. A good experiment is to beam a low-cost laser down onto a tilted glass slide, and see that two laser point images are made: one that passes through the slide, and a second that is reflected by the slide. You can see how rotating the slide causes the reflected image to rotate.

I may be wrong on one or more of these explanations, but who can doubt that there is not a perfectly fine particle reflection explanation for refraction or any other phenomenon of light? Certainly D2B owners and probably D2B consumers already know all this and much, much more, but aren't telling the poor D2B excluded anything.

The neutron is probably a hydrogen atom

Just like the bizarre D2B rule that nobody can reveal how all matter is made of light particles, so it may be that another minor misleading claim is that the neutron is not just a hydrogen atom. James Chadwick, the person who first named the neutron supposed the neutron to "consist of a proton and an electron in close combination" with a mass "slightly less than the mass of the hydrogen atom"[56,57] If neutrons are actually Hydrogen atoms, then so-called neutrons could be combusted with Oxygen just like Hydrogen atoms. Certainly a neutron is made of light particles, and perhaps the emission and absorption spectrum are the same as that of Hydrogen.

---

[56] J. Chadwick, "Possible Existence of a Neutron", Nature, vol 129, 1932, p312.
http://www.nature.com/nature/journal/v129/n3252/pdf/129312a0.pdf

[57] J. Chadwick, "The Existence of a Neutron", Proceedings of the Royal Society of London. Series A, Containing Papers of a Mathematical and Physical Character, Vol. 136, No. 830 (Jun. 1, 1932), pp. 692-708.
http://www.jstor.org/stable/95816

Why no telling the public about the details of evolution, the history of science, and the future?

Where is the movie "Evolution" for the large screen? How about the movie "Science"? And then the most interesting movie of all: "The Future" – where are they? Again, this is a case where the silence is one of the best pieces of evidence of some kind of large scale conspiracy of silence - of creating a Pol-Pot kind of society where a group of elites sees and knows everything, while purposely leaving the public in absolute ignorance and misleading them with obviously false "official" stories and explanations.

I have spent the last 8 years researching the history of science. Making my "Universe, Life, Science, Future" database and videos has been a full time hobby that I have worked consistently on for all that time. So it gave me an unusually good perspective on the course of science history as experienced and recorded by many of the great scientists of the past.

You owe it to yourself to see my excluded version of the complete story of the universe, the evolution of life, the history of science, and a projection into the far future all in a 10 minute free movie at ulsfmovie.org. The other progressively longer versions have far more details and important information, also for free.

"Non-Euclidean" surface geometry unlikely to apply to the universe

The origins of so-called non-Euclidean geometry start with the Russian mathematician Nikolay Lobachevsky, who, in 1829, was the first to publish a non-Euclidean geometry[58]. The goal of "non-Euclidean" geometry is to disprove one or more of Euclid's postulates, and what grew out of this effort was the rise in the popularity of the mathematical field of *curved surface geometry*.

Lobachevsky published the first known instance of the famous claim that a triangle made of curved lines may have angles that add to more than 180 degrees (for example on the surface of a sphere), an apparent violation of the Euclidian postulate that all angles of a triangle must add to 180 degrees (or pi radians). Another case is when the angles of a triangle add to less than 180 degrees (for example on the surface of a hyperboloid). One problem with this is that there is not a single angle formed between two curved lines (even on a plane); the angle changes the closer to the intersection one measures.

This new idea of restricting space to a curved surface was then applied to space in the universe. The argument was that space in the universe may be "curved", but like looking at a small part of a very large curve, only appears to be straight. There were early critics of this new and unlikely idea that space in the universe might be somehow "curved", including the very influential Hermann Helmholtz[59], which I hope to describe in more detail in the future.

[58] NI Lobachevsky, "On the foundations of Geometry", Kazan Messenger, 1829.

[59] Helmholtz, H., "Über die tatsächlichen Grundlagen der Geometrie", Verhandlungen des naturhistorisch-medicinischen Vereins zu

Even if we accept that there are some shapes and/or spaces that are not described by, or that violate one or more of Euclid's postulates, and so can be called "non-Euclidean", there is a simple truth that any "non-Euclidean" surface geometry is only a subset of unrestricted space. What is being called "non-Euclidean" geometry is more accurately called "curved surface geometry", because all that this geometry does is to restrict the possible values of the variables that represent position (traditionally for 4 dimensions the variables x, y, z, and t, or for two dimensions u and v). The possible positions are often limited to the surface of some geometrical shape like a sphere, or hyperboloid. So space in a surface geometry is a *subset* of unrestricted space, and I think that the traditional unrestricted three dimensional space is the most accurate model of space for the universe.

To D2BW excluded ears, it may sound overly conservative, but in my opinion (and no doubt in the opinion of those receivers of D2BW who control many micro and nano scale particle devices), the traditional view of unrestricted space and time as described by four variables (x,y,z,t) is not only more accurate and logical than restricting space and time to a curved surface geometry, but is far more simple.

Heidelberg, 4, 1866, pp. 197-202.
http://books.google.com/books?id=hksDAAAAYAAJ&pg=PA197
English: "On the Actual Foundations of Geometry"
Some parts translated in:
Joan L. Richards, "The Evolution of Empiricism: Hermann von Helmholtz and the Foundations of Geometry", Brit. J. Phil. Sci. a8 (1977), p235-253.
http://www.jstor.org/stable/686808

One theory I am putting forward here is that this new abstract curved-surface geometry was and still is embraced and funded by many D2BW owners and consumers (who don't make any use of it themselves) because it helps to distract and keep excluded people away from understanding and being interested in science.

The special and general theories of relativity are very unlikely

This is not to say that I reject the claim that nothing can move faster than the speed of light, or that gravitational and inertial acceleration are not equivalent, but just that I reject 1) that light is massless, 2) that the speed of light is always constant, 3) that mass and motion can be changed into each other as implied by equations of momentum and energy, 4) the idea that space and or time contracts or dilates depending on the motion of matter, and 5) that a surface geometry applies to space and time in the universe.

I reject the theory that light is massless in favor of light being made of material particles which are the basis of all matter (as the smallest known particle). This is obvious when we see light emitting from a burning candle and the candle made smaller in mass as a result. So I define the label "photon", not as a quantum of energy, but as a material light particle which is the basis of all matter.

I think that light particles can reflect off of each other, and the evidence of this is simply that light reflects off of objects. So for light to reflect, it seems likely that the velocity of light (which is simply space covered over time) can not only be changed, but can be 0. We can imagine a particle of light reflecting off of other particles in a space that becomes filled with more and more light particles. As the number of light particles in the space increases, the space and time between collisions decreases, and so eventually, a light particle may be held motionless, packed together with many other light particles. In fact, we may be sitting on an atom factory. Atoms may form near the surface of planets and stars where space becomes less dense and there is room for atoms to form without being torn apart by collision. In 1940, Haxby et al at Westinghouse showed that even light particles can cause atomic fission[60]. So like molecules, atoms too, may be objects that can only exist at relatively low temperature and density.

Many people hold up the chain-reaction of uranium fission, and the equation $E=mc^2$ as proof of the theories of relativity and the equivalence of energy and matter, but atomic fission was the result of experiments done by Enrico Fermi, Otto Hahn and Lise Meitner, not the result of trying to confirm a physics theory like relativity. In addition chain-reactions can be just as easily explained with a particle collision chain-reaction theory, and as I explained before, any equation of energy or momentum presumes that mass and velocity can be exchanged. While this matter and motion interchangeability may be a useful generalization to describe some physical phenomena, it seems, in my mind, unlikely that motion can spontaneously change into matter, or matter spontaneously change into motion; I find more likely the theory that motion is simply *transferred* between pieces of matter by collision. It may seem hard to believe that a small motion can lead to an apparently much larger motion, but typical combustion (burning of hydrocarbon and oxygen) involves the same exact phenomenon- a small motion in for a large and sometimes long duration motion out. One example is making an opening in a container with a material under high pressure; a little motion in results in much more motion out. In this view, light particles are trapped inside some high pressure material container within atoms, and when this container is broken, the light particles fly out. Another theory is that light particles have a lot of motion within atoms,

[60] R. O. Haxby, W. E. Shoupp, W. E. Stephens, and W. H. Wells, "Photo-Fission of Uranium and Thorium, Phys. Rev. 58, 92-92 (1940). http://prola.aps.org/abstract/PR/v58/i1/p92_1

but that this motion is directed in a closed orbit within the atom, and chain reactions simply knock them out of orbit so they fly out in all directions. Either way, I think that all chain reactions like atomic fission and combustion involve the same simple process of atoms separating into their source light particles. One classic defense of the theories of relativity is that we can't add the speed of moving objects to the speed of light emitted or reflected from them. But light particle collisions with or within an object may be so short that the added motion of the much larger object in that instant of collision is nearly zero. In addition, light particles inside an object may not be physically connected to it, and so may not take part in the object's collective motion. One big question that modern people do not often address is where all the light particles in an exothermic reaction (like an atomic fission chain reaction, or simple combustion reaction) come from. The current view simply waves them away as "energy", there is no quantity of "photons" on either side of any chemical equation. For combustion reactions, the current explanation is that the light particles emitted are lost from electrons. But I think it seems likely that entire atoms (including protons and "neutrons") may be separated into their source light particles in combustion and other exothermic reactions. If true perhaps some apparently non-atomic exothermic (or "exophotonic") reactions might result in radioactive atoms, different atoms, or different atomic isotope products. Probably D2B owners already know the answers to these questions, but like D2BW, are not sharing them with the public.

In addition I reject the theories of relativity because I reject the theory of space and time dilation and contraction, and I reject the theory that time depends on the motion of matter. For example, a clock may move more slowly under water, but that doesn't mean that time in the universe is slower, or that time as experienced by the clock is slower. I think that there is just one time for the entire universe. It seems logical to me that if the time is 10 o'clock here, it is also 10 o'clock in every other galaxy. Herbert Dingle identified a simple problem with the so-called "twin paradox" claim of time dilation in which one twin ages more slowly because they are moving faster than the other twin: how could one twin be moving at a different velocity compared to the other twin when their motion is relative to each other?[61]

There are a number of reasons why the general theory of relativity is popular. I think that the number one reason is that those who own and control remote neuron writing, as crazy as it sounds, want to keep the public as far away from understanding the details of seeing, hearing and sending thought images and sounds as possible- they simply don't want any competition, or for the public to see all the past images of many unpunished violent crimes. Secondly, very few people have ever learned calculus or how math is used to determine future locations of pieces of matter. Perhaps another reason for the popularity is that many people think that a new theory is an improved theory, but just because a theory is new doesn't mean that it is a better theory. There are plenty of examples where the new theory was certainly no better and many times much worse than the earlier most popular theory: the rise of Christianity and Islam in the Mediterranean replacing Polytheism around 400 and 700 AD respectively, the rise of Nazism and fascism in parts of Europe during the 1930s, etc. In addition, Einstein was a colorful character, and represented, to many people, the opposition to the rising Nazi movement in Europe. Perhaps Einstein's theory winning, represented the anti-Nazis winning in the

---

[61] Herbert Dingle, "Relativity and Space Travel", Nature, **177**, 782 (1956).

minds of many people, and so they agreed to support these unlikely theories. Many Nazi's famously rejected Einstein's theory of Relativity and so opposition to Relativity, became associated with support for the Nazis, even though much of the opposition to theories like relativity are from a desire for truth and accuracy, and from a fear of inaccurate claims becoming popular.

There have been many famous critics of Einstein's theories of relativity, some memorable critics are Charles Lane Poor[62], William Pickering[63], and Herbert Dingle[64].

One related and relevant question of these centuries is: "Did the person receive direct-to-brain windows or were they one of the excluded?". For example, did Isaac Newton get D2B? Did Beethoven? Did Ben Franklin? Did Abe Lincoln? Did Einstein? Did JFK? Did Marilyn Monroe? Did Elvis? Either way, the truth about D2B significantly changes what we know from the public story. They either watched videos in their eyes, or knew nothing about it, and were the victimized excluded. Did Einstein get D2B? It seems likely to me that he did, and compared to relativity, that theory is probably the more accurate one.

One of the saddest parts about the rise of non-Euclidean geometry and the theory of relativity is that it has been one of the best and most successful efforts against arousing public interest in science in years. Average people are shushed away from the big curtain of science and technology with the

---

[62] Charles Lane Poor, "Gravitation Versus Relativity", Putnam, 1922. http://archive.org/details/gravitationversu00poorrich

[63] Pickering, W. H., "Shall We Accept Relativity?", Popular Astronomy, Vol. 30, 04/1922, p.199. http://articles.adsabs.harvard.edu/cgi-bin/nph-iarticle_query?bibcode=1922PA.....30..199P

[64] Herbert Dingle, "Science at the Crossroads", Martin, Brian and O'Keefe Ltd, 1972.

excuse that they are not qualified to understand the universe, science and technology, when the exact opposite is true. Most of the universe is very simple and logical in my opinion. For example, that all matter is made of light particles – how much more simple could it be? That our fate is to build a globular cluster, that thought images and sounds can be seen, heard and sent, etc. This reign of non-Euclidean geometry and relativity has been a century of dogma, but is small when compared to the millennia of religious dogma.

In my view, the much more likely theory is the "billiard-ball" universe, where light particles just move around the universe colliding into each other. Even the phenomenon of gravity, as we experience it, being pushed down to the surface of the Earth, may be the result of many particles colliding with us, particularly from above us. If gravity is just the result of particle collisions in a dense matter field, then any "anti-gravity" would be doubtful. The only way to move away from a dense matter field would still be the only method we know; to use material particle collisions to push our way out. It's impossible to sum up the positions and motions of millions of particles, but overall large-scale motions can be generalized using the simple equation for the force (or collective effect) of gravity. Simply using "iteration" (modeling a universe of many particles and measuring their mutual gravitational influence on each other for each frame of time) using Newton's simple $A_2 = Gm_1/r_2$ equation, and including collisions, is probably the most simple, fastest, and accurate method to model the universe. Almost certainly that, and the simple $F=ma$ inertial motion law are what the neuron owners have used for centuries when, for example, trying to determine how contracting a muscle will cause a body to move in the future.

The truth is that history is filled with millions of "mistaken beliefs" and just simply "lies". There is not space to list them all, but some are: the claims that the Sun goes around the Earth, of Angels, of a Heaven, of Demons, ghosts, spirits, witches, that Jesus rose from the dead, Moses parted the Sea, Lee Oswald killed JFK, Islamic hijackers brought down the World Trade Centers, that people didn't figure out how to see and hear thoughts, and we can add to the garbage bin of mistaken beliefs: the Big Bang expanding universe theory, background radiation, and both the special and general theory of relativity.

So what is the most accurate interpretation of the universe in my view?

With so many lies, frauds and mistaken beliefs, what is the average person to believe are the most accurate theories? First, I am sure, that those people who own D2B and many D2B consumers must have a much better understanding of the universe than the public is being shown. For myself, as I stated above, I support an "all inertial" "billiard ball" universe, where matter and motion are never created or destroyed, and are always conserved, but I think clearly that Newton's law of gravitation is the best and most useful equation to use when modeling matter in the universe. But even Newton's simple law may not help us to predict what living objects in the universe may do in the future. As I say above, clearly the theories of relativity, of time and space contraction and dilation, the expanding universe, the electromagnetic wave theory of light, and most of the claims of psychology and religions are all false. Instead, all matter being made of light, which are material particles in a universe of unrestricted space and time, is the most accurate truth in my opinion. Beyond that, the theory of a single common ancestor for all of life on Earth and of natural selection (the theory of evolution) is one of the few theories that is accurate and will survive the truth about RNRAW. One theory that I find impressive is that, since the universe is probably infinite is time and size, perhaps at the scale of galactic clusters, or at the atomic scale, there are collective objects that we would compare with ourselves as living objects.

I'm only interested in the truth and telling everybody else the truth, I'm not interested in lying to and tricking people. Another point is that I am not looking to ridicule those who believe religions and other inaccurate theories, but instead trying to win them over to what I think is the more accurate truth. I feel sorry for many people, because just like me, starting from birth, they have been tricked, lied to, duped, fooled with the lies and unlikely claims of religions, and by deliberately dishonest scientific claims such as the expanding universe theory, that light is not the most basic atom, that people haven't figured out remote neuron reading and writing long ago in the past, etc.

Timeline of Science

Here is a timeline that just highlights a very few important science events.

1208 CE

Figure 10. Robert Grosseteste claimed that all matter is made of light in 1208.

Robert Grosseteste (fig. 10) writes that all matter is made of light.[65] This is around the time of the first Universities (Paris, Oxford, etc.) in Europe, and is evidence of a collective exploration of science which, for smart people, would quickly focus on light particles, communication, electricity, micro-machining, motors, cameras, and neurons.

---

[65] Robert Grosseteste, "De Luce", 1208.

<u>1600</u>

Figure 11. Giordano Bruno, victim of the war on truth and science.

Giordano Bruno (fig. 11) is murdered, burned at the stake, in part, for his view of a moving Earth. This murder of scientists, engineers, teachers, those who tell the truth, and non-religious people, has been the theme on Earth for millennia and continues into the 2000s, but now the murdering and torturing is mostly done remotely with microscopic particle devices.

1664

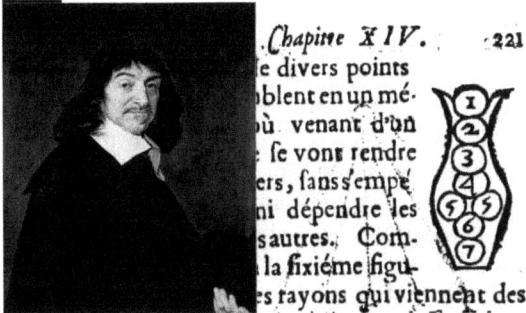

Figure 12. Image of René Descartes and a figure from "Le Monde".

René Descartes' "Le Monde" (fig. 12) identifies the two major theories for light, the wave and corpuscular theory[66].

1672

Figure 13. Isaac Newton, and figure from his first letter to the Royal Society.

---

[66] Descartes, R. Le Monde ... Ou Le Traité De La Lumière Et Des Autres Objets Principaux Des Sens, Avec Un Discours De L'action Des Corps Et Un Autre Des Fièvres, Composez Selon Les Principes Du Même Auteur. Michel Bobin et Nic. le Gras, 1664.
http://books.google.com/books?id=DHEPAAAAQAAJ
English translation: Rene Descartes, Translated by Michael S. Mahoney, "The World or Treatise on Light", Chapters 13 and 14.
http://www.princeton.edu/~hos/mike/texts/descartes/world/worldfr.htm

Isaac Newton (fig. 13) more clearly and firmly establishes a "corpuscular" theory for light.[67] The most obvious view that light is made of particles will collapse in the beginning of the 1800s (because of the rise in popularity of the transverse wave theory for light of Thomas Young). But a particle theory for light will be partially revived for a third time by Planck and Einstein in the beginning of the 1900s. Apparently it may be that when the century turns, every one hundred years, at least one significant science contribution is released to the public by the D2BW owners. Certainly the D2BW owners play a large part of the suppression of the simple "light is made of particles" truth to maintain their ridiculous 700-year head-start and monopoly of light particle technology.

1678

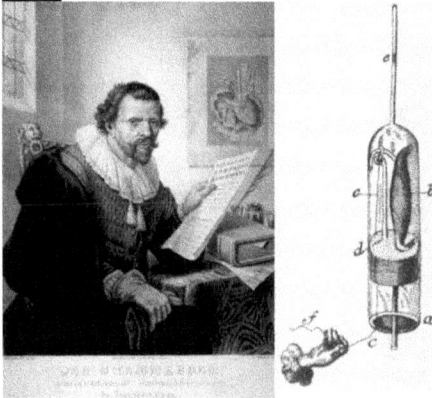

Figure 14. Image of Jan Swammerdam and drawing of his 1678 work.

---

[67] Isaac Newton, "Draft of 'A Theory Concerning Light and Colors'", Feb 6, 1671/2, in English, c. 5,137 words, 14pp. Shelfmark: MS Add. 3970.3, ff.460-466 Location: Cambridge University Library, Cambridge, UK
http://www.newtonproject.sussex.ac.uk/view/texts/normalized/NATP00003

The earliest publicly known direct neuron writing: Jan Swammerdam contracts a frog leg muscle using two different metals (fig. 14).[68] Most people are not aware of how far back into the past neuron writing actually goes. It simply was not called "neuron writing".

It may be that the exponentially growing, well organized government, military, communications and university secret research into remote neuron reading and writing (none of which gets published) is highly developed by this time (the late 1600s). If true, then almost all science here and later: Galvani, Joe Henry, Heinrich Hertz, etc. is actually probably 1) D2BW consumers publishing ("leaking") old findings, or 2) those denied D2BW re-inventing secret findings of the past.

---

[68] John Joseph Fahie, "A History of Electric Telegraphy, to the Year 1837", E. & F. N. Spon, 1884.
http://books.google.com/books?id=0Mo3AAAAMAAJ

1791

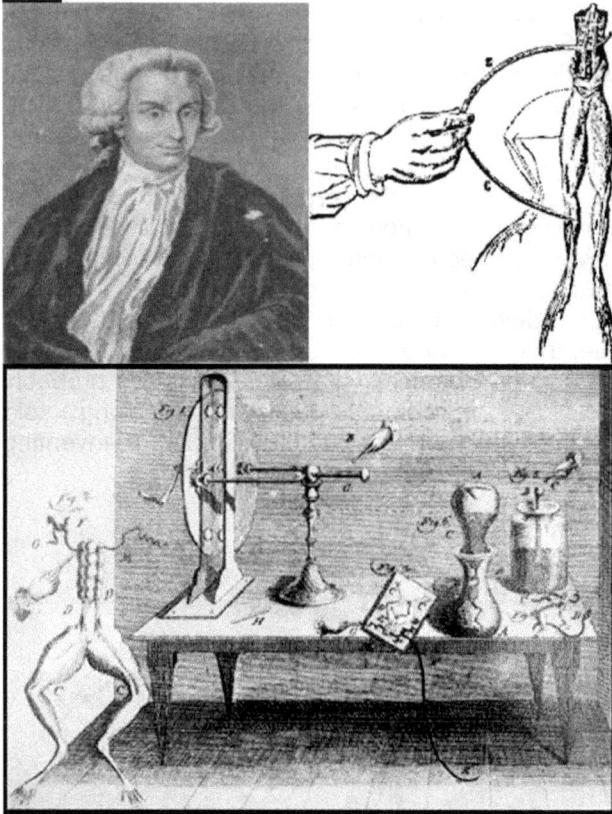

Figure 15. Portrait of Galvani and images from his 1791 paper.

Remote neuron writing[69]: Luigi Galvani makes a frog leg muscle move by touching the frog nerve with a scalpel while an assistant cranks a remote spark generator (fig. 15). Light from the spark reaches the scalpel, takes the form of electricity (photoelectric effect) and causes the frog leg to

---

[69] Luigi Galvani, Elizabeth Licht, Robert Green, "Commentary on the Effect of Electricity on Muscular Motion", Waverly Press, 1953.

contract. This is the first example of using light particles to contract a muscle remotely- all the way back in 1791. How many people know that remote neuron writing goes back at least to 1791? Isn't that something very important that we should know? Galvani also put the frog leg in between one piece of copper and one piece of tin to make the muscle contract. This will lead to the first electric battery. This 1791 publication is a clear indication that remote neuron activation may be widely known, although secretly, by this and all later times. The word "galvanized", taken from Galvani's last name, has come to have multiple meanings, for example "to be strongly set in opinion", but one secret meaning is that when somebody is murdered using remote neuron writing (for example to contract their lung muscles, or their heart) people say that they were "galvanized".

So you can say without any hesitation to anybody that "remote neuron writing goes back at least to 1791", because that is a clear and published fact.

1800

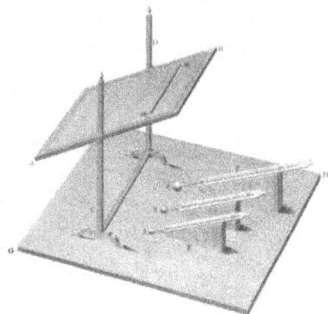

Figure 16. Portrait of Herschel and image from Herschel's 1800 paper (note the three thermometers).

Invisible light: William Herschel (fig. 16) reports that an invisible part of the spectrum (the infrared) heats a thermometer more than any other part of the spectrum.[70]

1801

The absolute length and frequency of each vibration is expressed in the table; supposing light to travel in $8\frac{1}{2}$ minutes 500,000,000000 feet.

Colours.	Length of an Undulation in parts of an Inch, in Air.	Number of Undulations in an Inch.	Number of Undulations in a Second.
Extreme -	.0000966	37640	463 millions of millions
Red - -	.0000256	39180	482
Intermediate	.0000246	40720	501
Orange - -	.0000240	41610	512
Intermediate	.0000235	42510	523
Yellow -	.0000227	44000	542
Intermediate	.0000219	45600	561 (= a⁴⁸ nearly)
Green - -	.0000211	47460	584
Intermediate	.0000203	49320	607
Blue - -	.0000196	51110	629
Intermediate	.0000189	52910	652
Indigo - -	.0000185	54070	665
Intermediate	.0000181	55240	680
Violet - -	.0000174	57490	707
Extreme - -	.0000167	59750	735

Scholium. It was not till I had satisfied myself respecting all these phenomena, that I found in Hooke's Micrographia, a pas-

Figure 17. Thomas Young and his table showing the wavelengths (particle spacings) and frequencies for various colors of light.

Thomas Young (fig. 17) shows that color is related to the frequency of light, and measures the wavelengths (particle spacings) for different colors.[71] In determining frequencies for light Young made a valuable contribution to human progress, but his opposition to a particle theory for light helped to set back the human species for more than a hundred years.

---

[70] William Herschel, "Investigation of the Powers of the Prismatic Colours to Heat and Illuminate Objects; With Remarks, That Prove the Different Refrangibility of Radiant Heat. To Which is Added, an Inquiry into the Method of Viewing the Sun Advantageously, with Telescopes of Large Apertures and High Magnifying Powers.", *Philosophical Transactions of the Royal Society of London* , Vol. 90, (1800), pp. 255-283.

http://books.google.com/books?id=dlFFAAAAcAAJ&pg=PA255

[71] Thomas Young, "The Bakerian Lecture: On the Theory of Light and Colours", Philosophical Transactions of the Royal Society of London (1776-1886),Volume 92, (1802), pp12-48.

http://journals.royalsociety.org/content/q3r7063hh2281211/?p=422e575 bae414c9a974a16d595c628d0ÏÇ=24

1816

Figure 18. Joseph Niépce and the first public photograph.

First photograph: Joseph Niépce makes the first publicly known photograph. The above image (fig. 18) on the right is from 1826 by Niépce (on the left) and is the first publicly known permanent photograph. But clearly the microscopic wireless floating and flying cameras must have been developed very early – perhaps even by 1200 as absurd as that sounds.

1827

Figure 19. André-Marie Ampère.

In the famous paper by André-Marie Ampère (fig. 19), which describes how two current carrying wires move toward or away from each other depending on the direction of the current, one paragraph contains a sentence using the words "suggère" (suggest) and "contractions musculaires" (muscle contractions).[72] The idea of using electricity to contract muscles also may imply the secret research of artificial muscles and walking robots that move much like humans by this time. An artificial muscle could easily be constructed using this simple phenomenon by drawing wires together to contract a flexible rubber material.

Félix Savary describes the phenomenon of electrical oscillation[73]. How an electric current oscillates between a capacitor (a Leyden jar) and an inductor. This is a basic part of wireless communication.

---

[72] André-Marie Ampère, "Théorie des phénomènes électro-dynamiques, uniquement déduite de l'expérience. ", Méquignon-Marvis, 1826
http://www.ampere.cnrs.fr/ice/ice_math.php?typebookDes=Oeuvres&bdd=ampere&bookId=23
http://gallica.bnf.fr/ark:/12148/bpt6k29046v
A partial English translation is in:
Tricker, R. A. R., "Early Electrodynamics - The First Law of Circulation", (Pergamon, NY), 1965, p155-200.
[73] Félix Savary, "Mémoire sur l'alimentation", Annales de Chimie et de Physique, 1827, 34:54-56.
http://books.google.com/books?id=QaQwAAAAYAAJ&pg=PA54

1842

Figure 20. Joseph Henry.

Joseph Henry (fig. 20) describes the basis of radio when he reports magnetizing a needle that is "...7 or 8 miles away..." by electrical induction from an electric spark.[74]

[74] Joseph Henry, "On Induction from Ordinary Electricity; and on the Oscillatory Discharge.", Proceedings of the American Philosophical Society, vol. II, 1842, p193-196.
http://books.google.com/books?id=5AIwAAAAIAAJ

1861
October 26

Figure 21. Philip Reiss

Philip Reiss (fig. 21) goes public with the first known microphone, telephone and speaker. Sound can now be converted to electricity and back to sound again.[75] Quietly sending sounds over long distances is now possible. Reiss dies at a young age and may have been remotely murdered; made an example of, symbolically, for going public with encoding and decoding sound in electricity (the telephone).

1864 CE
James   Clerk   Maxwell   publishes   his Electromagnetic theory of light[76], shockingly, still

---

[75] Silvanus Phillips Thompson, "Philipp Reis: inventor of the telephone: A biographical sketch, with ...", 1883.
http://books.google.com/books?id=7uQOAAAAYAAJ
[76] James Clerk Maxwell, "A Dynamical Theory of the Electromagnetic Field", Royal Society Transactions, Vol. 155, 1865, p. 459-512.
http://books.google.com/books?id=xVNFAAAAcAAJ&pg=PA459

currently widely taught and accepted as the most accurate theory of the physical composition of light.

1881

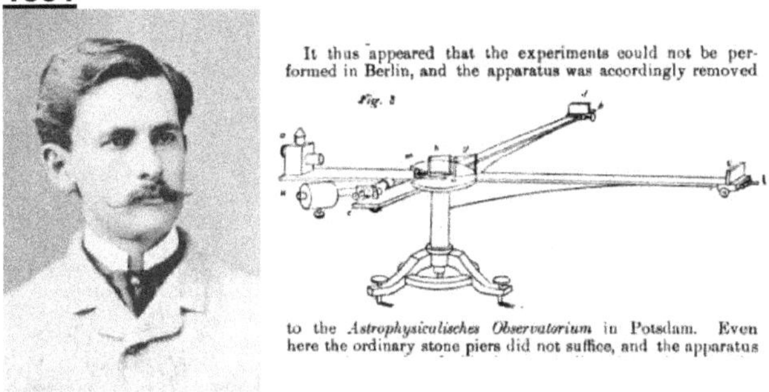

It thus appeared that the experiments could not be performed in Berlin, and the apparatus was accordingly removed

to the *Astrophysikalisches Observatorium* in Potsdam. Even here the ordinary stone piers did not suffice, and the apparatus

Figure 22. Albert Michelson and a drawing of the apparatus he used to measure and compare the speed of light in two directions at the same time.

Albert Michelson with Alexander Graham Bell's support initiates the end of the "light is a wave in an aether medium" theory by showing that light travels at the same speed with no change due to the movement of Earth in the supposed aetherial medium (fig. 22)[77]. Albert Einstein reinforces this end of aether in 1905.

1885

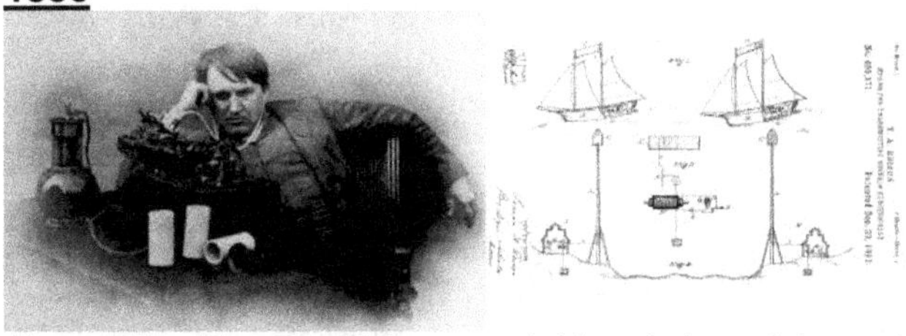

Figure 23. Thomas Edison and his wireless telegraph patent.

---

[77] Albert A. Michelson, "The relative motion of the Earth and the Luminiferous ether", The American Journal of Science, Volume 122, 1881, p120.
http://books.google.com/books?id=S_kQAAAAIAAJ

Invisible particle communication. Thomas Edison describes the first publicly known sending and receiving of text messages by invisible frequencies of light particles (wireless) (fig. 23).[78]

<u>1887</u>

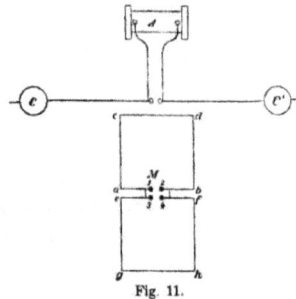

Figure 24. Heinrich Hertz and a drawing from his paper.

Heinrich Hertz publicly explains "electrical resonance" (which allows specific ranges of frequencies of light particle beams to be filtered) (fig. 24).[79] This greatly popularizes the idea of low frequency particle (wireless) communication. Many people wrongly credit Hertz with inventing radio communication - Hertz's big contribution was going

---

[78] Edison patent 465,971, "Means for transmitting signals electrically". http://www.google.com/patents/US465971?printsec=drawing#v=onepage&q&f=false

[79] H. Hertz, "Ueber sehr schnelle electrische Schwingungen", Annalen der Physik, Volume 267 (V. 31) Issue 7, March 1887, Pages 421 - 448. http://books.google.com/books?id=WhY4AAAAMAAJ AND http://de.wikisource.org/wiki/Benutzer:CK85/Untersuchungen_%C3%BCber_die_Ausbreitung_der_elektrischen_Kraft_Kapitel_2 English Translation:
Heinrich Hertz, tr: D. E. Jones, "On Very Rapid Oscillations", Electric Waves, 1893, 1962, p29. http://books.google.com/books?id=EJdAAAAAIAAJ

public with tuning in specific frequencies by electrical resonance and publicizing radio as a method of communication. It seems likely that Hertz, like Philip Reiss, the first to go public with the telephone, may have been murdered remotely with particles or microscopic devices for his telling the public about secret technology.

1889

Figure 25. William Friese-Greene and the first (publicly known) films made on celluloid (1889-1890).

William Friese-Greene (fig. 25), invents the earliest public motion picture camera, describes capturing photographs from light emitted by the eye, and hypothesizes about capturing images from the eye from behind the eye.[80]

---

[80]   William Friese-Greene, "Photographs Made with the Eye", Photographic Times, 1889, p108-109.
http://books.google.com/books?id=-bUaAAAAYAAJ&pg=PA108
(and see the movie "The Magic Box")

1895

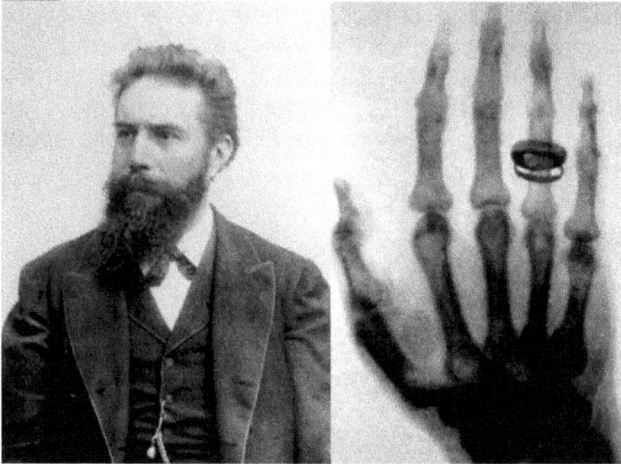

Figure 26. William Röntgen and an early X-ray photo.

Wilhelm Röntgen reports finding x-rays[81] (fig. 26).

---

[81] Wilhelm Conrad Röntgen, "Über eine neue Art von Strahlen", Aus den Sitzungsberichten der Würzburger Physik.-medic. Gesellschaft 1895.
http://de.wikisource.org/wiki/%C3%9Cber_eine_neue_Art_von_Strahl
English translation:
"On a New Kind of Rays", Nature, Volume 53, Number 1369, Jan. 23, 1896, p274.
http://books.google.com/books?id=nWojdmTmch0C&pg=PA274

Loches = 2 mm. *D* ein mit phosphorescirender Farbe über-
zogener Glimmerschirm. Die Glaswand *E* muss möglichst
gleichmässig und ohne Knoten, der phosphorescirende Schirm

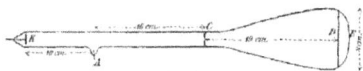

Fig. 1.

so angebracht sein, dass man durch das Glas und den Glimmer
hindurch den von den Kathodenstrahlen hervorgebrachten
Fluorescenzfleck sehen kann. — Für manche Versuche ist es
zweckmässig, den Glimmerschirm unter 45° gegen die Rohraxe

*Figure 27. The first public electric display and Ferdinand
Braun.*

The first electric display (cathode ray tube), by
Ferdinand Braun[82] (fig 27).

We can even play this timeline forward into the
future with expected technology going public or
being invented:

2014

1. **Outside glasses –**
   digital camera
2. **Inside glasses – eye**
   movement sensor will
   direct the camera
3. **Side of glasses –**
   digital processor and
   wireless transmitter
4. **Brain implant – small**
   implant under the skull
   will receive wireless
   signals and directly
   stimulate the brain's
   visual cortex

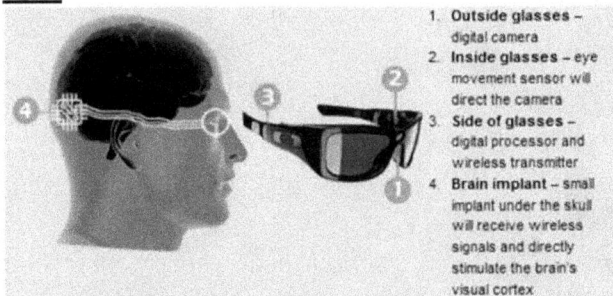

Figure 28. Monash University's "Direct to brain bionic eye"

Monash University plans to patent a "Direct to brain
bionic eye": "a small implant under the skull will
receive wireless signals and directly stimulate the
brain's visual cortex"[83] (fig. 28). This form of extra-

---

[82] Ferdinand Braun, "Ueber ein Verfahren zur Demonstration und zum
Studium des zeitlichen Verlaufes variabler Ströme", Annalen der Physik
und Chemie, vol. lx., 1897, p. 552-559.
http://books.google.com/books?id=rXgMAAAAYAAJ&pg=PA464
[83] http://www.monash.edu.au/bioniceye/

cellular electronic device assisted remote neuron writing will allow those without sight to see.

2015
Sound a brain hears recorded remotely.
Microscopic camera.

2018
Radio device functions as cell organelle.

2020

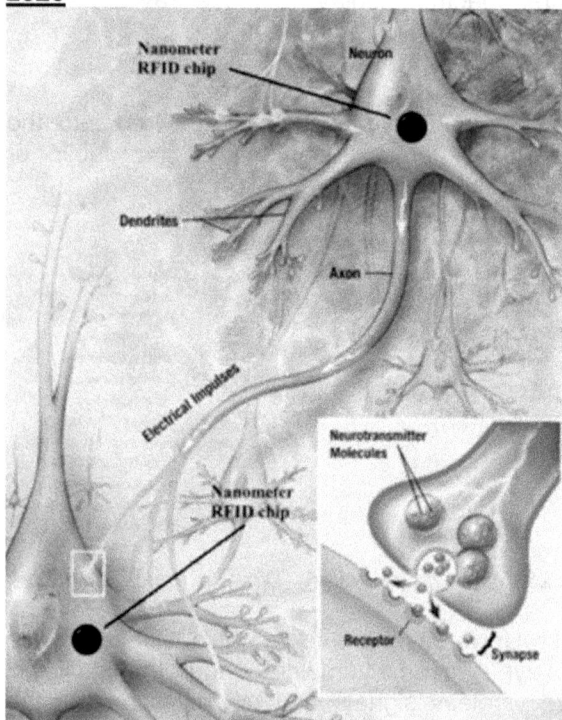

Figure 29. Perhaps what an RFID device may look like in a neuron.

Remote neuron writing using intracellular microscopic devices in neurons (fig. 29).

2025
Thought-images seen.
Thought-audio recorded and played out loud.
Humans start to communicate publicly by thought image and sound only.
Microscopic flying camera.

2040
Artificial muscle walking robot.

2050

Figure 30. Humans walk around with robot servants, notice the eye and thought screens and D2B windows.

Humans walk around with robot servants (fig. 30).

2100

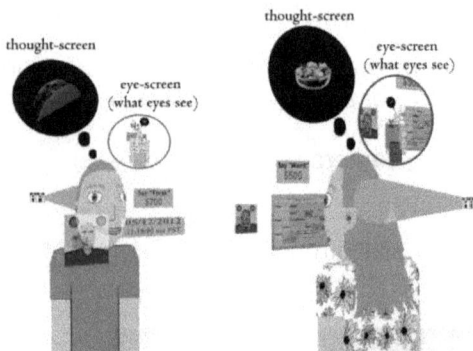

Figure 31. What humans communicating by thought using direct-to-brain windows may look like.

Most humans communicate only by images and sounds of thought (fig. 31).

Figure 32. Helicopters form lines of traffic above the street. Even large artificial muscle wing flapping flying vehicles are a possibility- the pterosaur "Quetzalcoatlus" shows that it's possible.

Helicopter-cars form a second line of traffic above the streets (fig. 32).

100 ships with humans orbit Earth.

2140

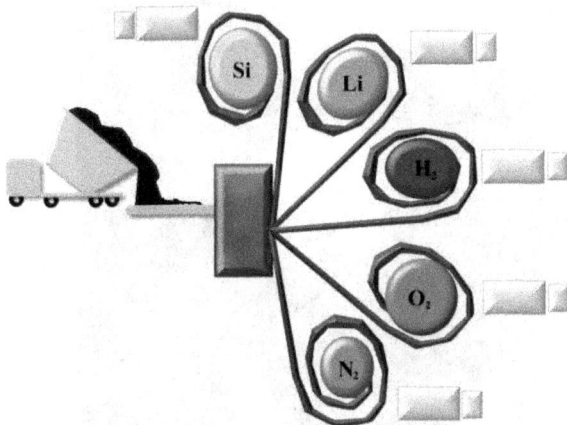

Figure 33. Large scale transmutation: waste goes into a particle collider, and charged atom fragments are separated by mass using an electric field.

Large scale transmutation (fig. 33): common atoms like Iron converted into Hydrogen and Oxygen using particle accelerators and colliders.

2220
Robots do most low-skill jobs.
1000 human-filled ships orbit earth.

2270
Humans live on Mars.

2500
End of death by aging. Growth development of a body can be made to go forward, stopped, and/or reversed by nanometer scale devices changing the order of DNA nucleotides.

<u>2550</u>
Humans live on Venus.

<u>2570</u>

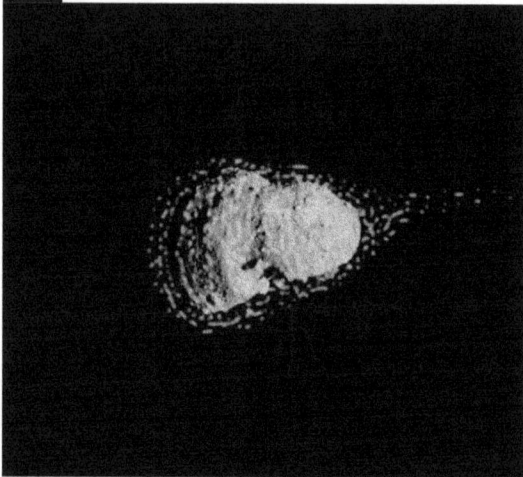

Figure 34. Humans will probably use the masses of many ships (gravity) and thrust to move large bodies around the star system.

Humans move an asteroid (fig. 34).

<u>2650</u>
Humans create atoms from light particles.

2750

Figure 35. The first ships to reach a different star will be an epochal moment for humans of Earth.

Ship reaches other star (Alpha Centauri) (fig. 35). First close up pictures of planets of a different star. Living objects found around another star (bacteria made of DNA found on planets of Alpha Centauri).

2800
Humans change the motion of a moon.

<u>2850</u>

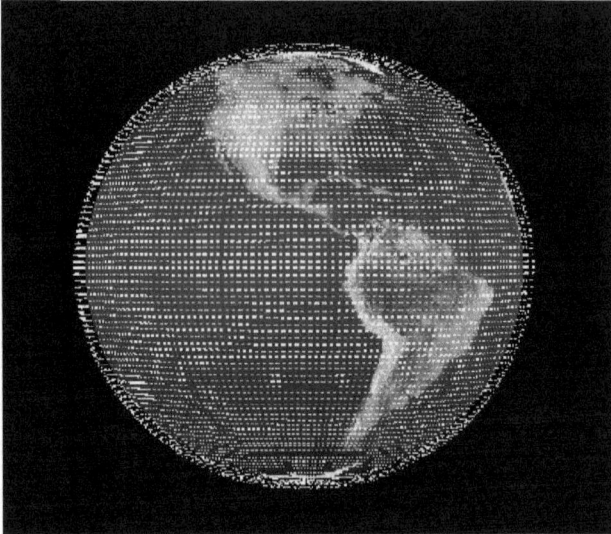

Figure 36. The Earth may look like a bee-hive of ships in the future.

Humans change the motion of a planet (the Earth) (fig. 36).

<u>2900</u>
Ship impacts surface of Jupiter. First image of surface of Jupiter.

<u>3200</u>
Ship from Centauri reaches Earth with objects.
Humans reach a different star, Centauri.

3500
Atmosphere of Venus removed.

4000

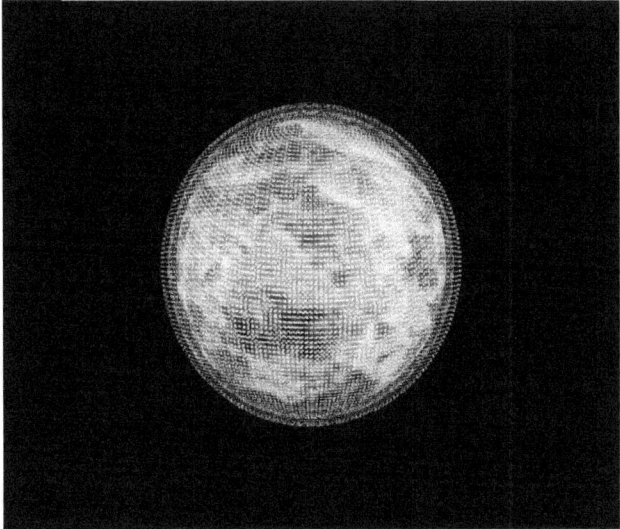

Figure 37. Ships consume all the atmosphere of Jupiter to reveal the massive liquid red-hot molten surface.

Atmosphere of Jupiter removed (fig. 37).
Humans have ships at 10 stars.

4500
Motion of all planets under control.
Humans reach center of Earth.
Humans live on Jupiter.

5100

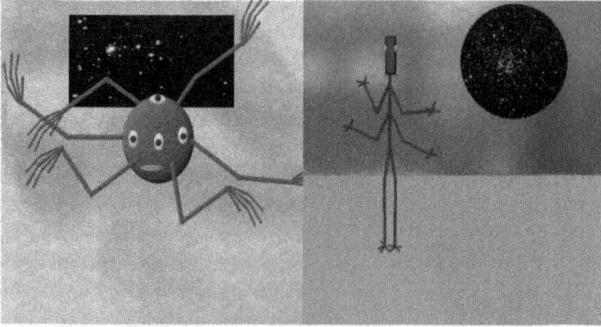

Figure 38. Probably some organisms will be adapted to low gravity, and others, like us in our current form, to higher gravity.

Image of advanced life of a different star (fig. 38).

5500

Figure 39. Thousands of ships move the Sun using only thrust and gravity, without ever needing to touch it.

Motion of star controlled (fig. 39). Earth star moved in direction of Centauri.

6000
Humans touch advanced life of a different star.

17000
One trillion humans.

27000

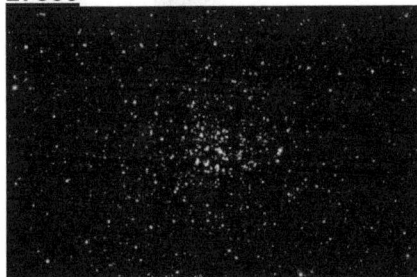

Figure 40. How the stars humans occupy might look in 25,000 years.

Humans inhabit 100 stars and form a globular cluster of 10 stars (fig. 40).

47000

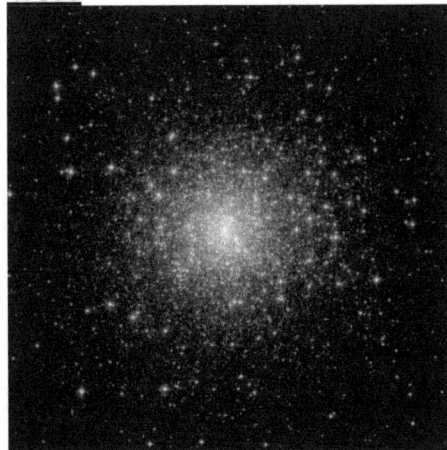

Figure 41. How the stars humans occupy might look in 45,000 years.

Humans inhabit 1000 stars and form a globular cluster of 100 stars (fig. 4.77).

65000

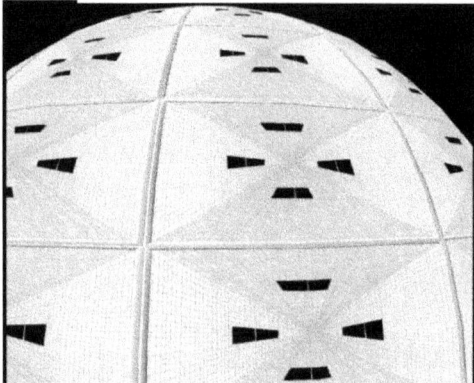

Figure 42 Earth is completely filled: "I live in unit 58173639203, enter in sector 25, take a left, and then a down."

Earth is completely filled with living objects (fig. 42).

130,000

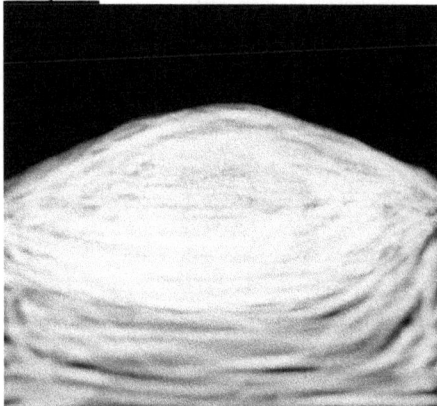

Figure 43. How the Sun might look in 128,000 years.

Star of Earth completely consumed by living objects (fig. 43).

20 billion years

Figure 44. How the Milky Way Galaxy may look in 20 billion years.

The Milky Way is by this time a Globular Galaxy (fig. 44). You would think that people in science would have told us about this likely future centuries before now, but no!

Other Important Ideas

Just like excluded people have not been told mind-numbingly simple science truths like that all matter is made of light particles, so they have not been told mind-numbingly simple and logical social ideas. Not everybody has to agree all the time on every issue. There is plenty of room for disagreement, debate, and new ideas.

First strike violence (FSV) is the big evil but somehow the public hasn't realized it

Violence is the big evil on Earth. Nonviolent activity (theft, trespassing, perjury, molestation, etc.) are a far lesser evil. But yet, the many murders of history and the apathy of the public to expose and try to stop them (in particular the most obvious ones like the JFK, RFK, and 9/11 murders) is evidence of this public's misplaced priorities. Unlike the victims of nonviolent crimes, the victims of FSV many times feel pain, suffer painful and lasting injuries, physical scars, and death. Many victims of violence are gone forever and aren't coming back, unlike victims of nonviolent crimes such as prostitution, recreational drug use, molestation, theft, etc. Nonviolent, although very annoying activities, like molestation, theft, lying, trespassing, copyright violation, etc. are not as serious as FSV activities. But much of the public has been hypnotized, or are slaves to tradition, in completely ignoring the most serious problem of first strike violence. For the current population of people, knowing who has sex-related crimes is a higher priority than knowing who has violent crimes. There is no "registry of violent offenders", and so as a result, people might hire, rent to, or date a person with multiple murders or assaults and have no idea, in particular if they are being excluded from D2BW. If forced to choose between spending a night locked in a room with a nonviolent offender or a violent offender, I think most logical people would choose the nonviolent offender.

Nonviolent activity that only hurts the self (drug and alcohol abuse, overeating) is a far lesser evil than FSV, but yet violent crime is of less concern than drug, pleasure and information related nonviolent "crimes". There are bizarre witch-trial-like sentences being handed out for those convicted with nonviolent crimes (drug use, pornography, molestation, copyright, violating national security, etc. offenses) even though the victims of the crime felt no pain and are living, while the victims of violent crimes felt a lot of pain and many are dead. Perhaps because they are dead and not around to seek justice, people tend to forget the crime. Perhaps nonviolent people are afraid to speak out against a violent person, because they fear becoming a victim of the violent person. Probably with the rise of science, logic, freedom of information, technology, and full and constant democracy where the public votes on all government decisions, the public will eventually realize that first strike violence is the biggest evil, and finally democratize, and create a logical punishment and sentencing structure, for the many various common crimes.

Perhaps many men may think that speaking out against violence is not manly, but stopping violence is the manly, courageous, and decent thing to do- it's staying silent that is cowardly and weak. We live in a time where fist fighting is obsolete, and people of any size or age can murder or seriously wound another person with the push of a button, and even remotely just with a "click" in thought. It's bizarre that hand guns are sold on the open market, but recreation drugs are illegal. Guns are used to cause far more damage, and D2B excluded people with a gun are a favorite target of D2B owners and consumers. The John Lennon, Giffords Safeway, Utøya Norway, and recent Colorado theater shootings and murders are all most likely classic examples of a D2B excluded hand gun owner being remotely controlled by remote neuron writing "suggestions" to do violence. Violence and murder done "indirectly" with an excluded, as opposed to directly with a particle device, and in particular, collectively spread among a large group of D2B conspirators and funders, creates more confusion about who to blame for a violent crime. The funders and executors of the neuron writing violent suggestions just cry "free speech" and say "we only showed them a picture of the crime...we didn't pull the trigger- they did!".

Just to make clear, I think that violence in self-defense from a violent attack is acceptable, of course, as most people can agree with. The long term desire though is to never have to resort to violence, by being able to stop violence before it occurs, which nanotechnology, remote neuron writing, robots, free information, and rapid full democracy is starting to enable. There are also complex examples where probably most people, myself included, would agree that violence is probably justified even though there is not an act of violence occurring, for example, somebody is pointing a loaded gun at another person, or where a person doing first-strike violence is imminent (within seconds). Many people believe that using violence (including murder) against a person who has broken into your house or apartment is justified. We should not forget that we all have loaded particle devices aimed at us all the time, so this fight against first strike violence is complex. Another example is firing back at those people who remotely molest people with particle devices; the molestation is non-violent, but it is annoying (making you think of a finger, contracting your muscles, making you itch, or to feel like you are getting an enema, etc.). When people write to your neurons, you can't write to your own neurons, and while it is nonviolent, it is very unpleasant and a violation of your body. The hope is for the remote molester's particle writing service to be stopped somehow, for a similar nonviolent but annoying molestation to be done back to them, and for all money involved, in addition to a fine ($100/second) to be transferred to the victim. There may be complex violence where many people vote and contribute money through their thought-screens to do violence, and a machine does the violence; with D2BW computers and nanotechnology, the list of conspirators and accessories of a violent crime can number in the thousands, but we need to expose, stop, and punish all of them.

I don't doubt that a "history of violence" movie has existed for a long time, although secretly, for D2B consumers- a movie that runs through all the major murders and shows clearly "who killed who". Such a movie should be assembled for the excluded public to reveal the truth about the many wrongly solved murders (certainly at least the most well-known of the wrongly solved murders, like those of JFK, RFK, and 9/11).

There is a simple truth: the more we expose, stop, and punish people who have done first strike violence (in particular those who remotely murder with particles, "galvanize"), the lower the chance there is of us being the victim of violence; the less we stop them, the higher the chance there is of us being the victim of violence.

Teach the public the details of evolution, the history of science, of the future, and of religions

I think that one reason people gravitate toward religions and creationism is simply because they have never been told and shown the history of evolution, science, and our possible future as owners and developers of a cluster of stars.

Most people know nothing or very little about the many details of evolution, for example that: 1) vertebrates (and insects) evolved internal fertilization (sexual intercourse) because on land (as opposed to in water) fertilizing the eggs directly had a selective advantage, 2) that one of our chordate worm-like ancestors evolved "upside down"- invertebrates like crustaceans and insects have their nerve chord near their front and not their back as we and the other chordates do, 3) that almost all fruits, nuts, and grains come from angiosperms, flowering plants (includes flowering trees), and that many fruit plants are closely related to each other- like "Fabales" (FƟBALEⱫ) the bean plants- most of the beans that we know are all closely related to each other; "Solanales" (SOLⱯNALEⱫ), the green pepper, tomato and potato are all relatives, and "Cucurbitales" (ꓘYUꓘRBⱵTALEⱫ)- cucumbers and squash are family. But flowers, which provide a large part of the food humans eat, only evolved around 150 million years ago, long after the first plant (1300 my {million years before now}), fungi (1200 my), animal (660 my), fish (560 my), insects (416 my), amphibians (375 my), reptiles (317 my), dinosaurs (228 my), and mammals (225 my). Birds (150 my) evolved around the same time as flowering plants. The list is very large, of very basic and interesting facts about the evolution of life on Earth that the public has not been shown and told on a large scale yet, like in a major motion picture called "Evolution", or as a mini-series like "Roots" on national television. The public has purposely been not shown and told about the history of science. Most people cannot name many major scientists of the past (such as Aristarchus, Eratosthenes, Galvani, Franklin, Volta, Descartes, Newton, Faraday, Fraunhofer, Michelson, Mendeleev, Edison, Leavitt, etc.), or those who led (and lead) the struggle for human rights, but can name many people in sports, acting and music. In particular, I think many people would benefit from learning that everything is made of light particles, the details of remote neuron reading and writing, and about the development of microscopic and nanometer scale cameras, transmitters, and flying devices. It's stupid and brutal to leave the public so terribly under informed and misled. Beyond that, many of the theories being called popular science are actually deliberate lies (like the expanding universe theory, that D2BW and RNRAW doesn't exist yet, that light is not material, that time changes based on speed, etc.). One myth being spread is that science is boring, but is that the truth, or just the voice in our head telling us that? In my opinion it's religion and most modern movies and television that are dull. I think the public would find very cute and funny recreations of the stories, for example, of Ctesibius playing the first organ, Vivaldi playing violin, William Hershel and his sister making lenses together for their telescopes, Trevithick driving the first steam carriage through town, the controversy of evolution, the women astronomers of Harvard; and find sad and tragic stories like the destruction of the Library of Alexandria (see the movie "Agora" for a recent re-enactment), the punishment of Galileo, the murder of Lavoisier, the burning of Joe Priestley's house, and smashing of the first spinning Jenny; and inspiring and emotionally stirring, how poor people like Michael

Faraday, Thomas Edison, Marie Curie, George Washington Carver, Elizabeth Blackwell, and many others used their genius to rise up and succeed in life through science; but also the very fascinating truths and inventions figured out, for example the first rockets, paper, and printing in China, Galvani and the electrified moving frog legs, how crowds gathered to see the first proof that the Earth rotates, shown by the change in direction of the Earth under a pendulum set in motion by Michael Foucault, the first x-ray photos of Röntgen, the first public telephone of Reiss, and the first motorized flight by the two Wright brothers against all the critics who said it was impossible. Part of the story of science are the big science lies and trivial pursuits, like the neuron reading and writing lie, the aether, the expanding universe, time dilation, black holes, background radiation, antimatter, nuclear forces, the neutrino, dark matter, the Higgs boson, etc. Here again, a major motion picture called "Science", or a mini-series on national television would help tremendously.

  Perhaps part of the story of science is the story of our future, which many insiders must know but are deliberately and callously keeping secret from the public. Probably first in importance in my mind is that our future is to build a globular cluster, and that there is clearly a pattern of galaxy formation, from nebula, to spiral, to globular galaxy. Of course, I think many people would be amazed and pleasantly surprised 1) to see what cities on the Moon and Mars will look like, 2) how the Earth will be a beehive of swarming ships, 3) that the atmosphere of Jupiter will probably be consumed by many ships to reveal a massive molten hot surface, 4) that walking robots will be doing all the labor for humans, 5) that massive scale atomic transmutation of common atoms into more useful atoms will probably be used to provide our descendants with water and fuel, 6) to know what the moment that the first ships that reach the second closest Sun to us, Alpha Centauri, might be like- maybe even tiny micro or nanometer scale cameras will be or have already been sent at high speeds to other stars using a particle accelerator, 7) to see that humans might actually reach the center of Earth and the other planets in the far future. Those are a few things I just thought of, there are many other interesting details about the future being denied. I can't believe that most excluded people would not be riveted by a major motion picture called "Future" that tells at least one of the many possible versions of this story, or perhaps a television mini-series on national television could enlighten a large majority of the public to these formerly secret great truths.

In addition to the history of evolution, science, and the future, much of the history of violence done in the name of religions is kept secret from the public, perhaps because the wealthy people who own much of the major media are religious, or fear that telling the truth may cause them to anger and lose income from religious people, but see books like "Holy Horrors" by James Haught[84] to get just a quick and minor introduction to some of the shocking and idiotic unprovoked violence done in the name of religions that much of the public never hear about. Here again, I think people on Earth would benefit significantly by seeing a history of the rise of religions and the brutality, conformity, violence, and destruction that has been done because of the many mistaken and radically inaccurate beliefs in religions. In particular the remote manipulation and tricking of many poor D2BW excluded people that have common mistaken religious beliefs, through remote neuron writing should be shown.

---

[84] Books by James Haught
http://www.amazon.com/James-A.-Haught/e/B000APE82Q

Full and constant democracy where people get to vote directly on the laws they must live under

One of the most important and most simple of the "social" ideas that the D2B excluded are being denied, is the idea of a full and constant democracy where the people get to vote directly on the laws they have to live under. This idea is extremely simple and fair. There are a wide variety of implementations and specifics, but generally it is majority rule: the laws with the largest vote and largest majority having higher importance than those with less votes. One idea is that the local majority may have a higher value locally than the global or multistellar majority, so that people with overly conservative or overly progressive views can individually shape their smaller planets, moons, and cities without everybody having identical laws and values.

Many people I tell this idea to dismiss it for a variety of reasons, two of the most common being: 1) the public isn't qualified to understand law, 2) they don't trust electronic voting. But a fully democratic system where the public votes directly over the i-net on all government decisions seems inevitable to me. You don't need to have any special law degree to understand the issues involved in voting "yes" or "no" to "do we send our kids to invade Iraq?", and other similar recent government decisions. In terms of trusting electronic voting, I have to point to the credit card system as a clear example of how electronic transactions are routine and not corrupt. We trust electronic transactions to buy and sell products with credit cards, but can't trust a similar system to record votes? We could even be voting on important decisions (like to invade a nation) by allowing people to record their vote on paper, but knowing that people have been reading and writing thought for 700 years, I think we can use the paperless system. It may be that some kind of system already exists, but only for those who receive D2BW. Perhaps those few D2B excluded who know their vote might be recorded vote in their thought audio or on their thought-screen. It may be that wealthy people "buy votes" in the D2B, and it is viewed as the free market. Would you vote for somebody if they paid you $1000 to? Many people probably would.

Full democracy is the next logical progressive step up after monarchy and representative democracy. In a monarchy, the public have no say over the laws. The next step up from there is a representative democracy, where the public can vote for representatives who create and vote on the laws. The final step is the transition to a full and constant democracy where the public create and vote directly on all the laws they are subjected to. Imagine how many people that lied and cheated their way into the Presidency would have been stopped early on, and how many bad laws and murderous decisions could have been overruled by the public had full democracy occurred many years earlier. The less people that can see and vote, the easier it is to corrupt, the more people that can see and vote, the harder it is to get away with a lie or unpopular decision.

We need to let the public vote directly on all laws to make sure that those laws do, in fact, still have popular support. It's unfair to subject people to a law that they don't get to vote on, or that doesn't have the support of at least a 51% majority of those who must obey the law. Maybe some of those laws held popular support at one time, but don't anymore. Similar to the complaint of the colonists in America who said that they were the victim of "legislation without representation", creating laws that the public is subject to but that they don't get to vote on is "legislation without participation".

Many people don't realize many of the interesting consequences and details of full and constant democracy, because we've never been told about such an obvious system as "full democracy" by our major media. There are many advantages. First the public can vote out bad government employees, can overrule unpopular Supreme Court decisions and any President's unpopular decision (like a pardon). There would be a big, although progressive upheaval. The majority would remove the many existing unpopular laws, combing out the current jungle of laws into a nicely trimmed garden, and new laws would rise that are the most popular and universally agreed on.   Even taxes, budgets, salaries, health care systems, etc. could be determined by letting the public vote.

Currently, we excluded can only imagine how much the D2B owners and secrecy influence and control who and what the representative governments do. This is one reason we need to open and democratize at least one major neuron and phone service for the public. There can be non-democratic neuron communication companies, but there should be at least one democratic choice so that the undemocratic telecoms are not the only people who maintain a system of electronic voting, and to decide who gets to see and hear thoughts.

Another interesting aspect of full democracy is that the military would be democratized. For example, generals would not be appointed, but would be elected. All the employments and salaries in the military could be democratically voted on.

Just like the laws, even the court system will probably be fully democratized. The jury will simply be those who record a vote, court decisions are made instantly, and solidified or overthrown as more and more people vote. Probably "military" courts will be replaced with one democratic court system and set of laws for all people of each nation and planet.

In addition to the laws, military, and courts, the other departments like police and fire departments can be democratically voted on so the public can decide who the best people for those departments are.

Another strong argument for shifting from a representative form of government to a democratic form of government is that the public, when allowed to vote on government decisions, will probably budget their precious tax money much better than representatives do. For example, look at what we get for the billions of dollars in income, sales, property, and other taxes: Free military "protection", police, fire-fighting, school, roads, food stamps for the very poor, social security after age 65, prison... and that's about it- no free food, free drinks, free health care, free clothes, free phone service (or free D2BW service- or one that pays us 1% of all profits made off of us), free transportation, free housing. I think many people can agree, that the public would probably promptly use those many billions in taxes to make their lives more comfortable just like the already ultra-wealthy representatives do. A trustworthy D2B consumer hinted that the votes of many, and perhaps even all, US representatives are routinely bought (probably through D2BW so there is no trace)- a typical example might be those Democrats who voted for an Iraq invasion under Bush jr. knowing that 9/11 was three controlled demolitions. In a free market even the votes of the public can be bought by wealthy people and organizations, so of course, even in a full democracy, wealthy people would always have a large influence on government decisions. Here's another one of a million examples of how we don't get our money's worth from a representative democracy: I made "ULSF" in 8 years in my spare time with no budget, but the United States Government National Science Foundation, whose job it is to promote science in the USA, and who currently have a yearly budget of $7 *billion* dollars, have not produced, to my knowledge, a single movie telling the basic story of evolution, of the history of science, or of the future of life of Earth. Like so much of "representative" government, where the public doesn't get to vote on the decisions, our money does not provide us with any free services. I mentioned above that our tax money does provide us with "free military protection", but much of this money is apparently being used for murders like those of 9/11, for galvanizations of innocent people, and for the constant remote particle assault and molestation that many tax payers feel all the time. Most of our money goes into lying to us, and keeping everything a secret from us. We and earlier generations of citizens paid for much of the microtechnology of D2B, but we are not even allowed to see the technology we bought, access the movie libraries, or even use it to communicate with each other- and then, for all that money we and many other hard working people paid- not only do we not get to use it, but we are constantly *abused* by it!

Some people argue that full and constant democracy is the equivalent to anarchy, but that is inaccurate, because there is still a government with a full democracy. The only difference is that with full democracy, government decisions are made by a few million people instead of by a few hundred people.

Full and constant democracy is the end product of that natural feeling that replaced monarchy with representative democracy: majority rule, the most good for the most people.

Ending forced labor

One obvious simple truth the public is not being told is that we should view military employment as a job, not as slave-labor or indentured servitude. The first change is to allow people in the military to quit without any punishment. We can't imagine people being jailed or fined for quitting employment from the police, or from a restaurant, why should working for the government be any different? I can foresee other future improvements to the military the public might vote up including ending hazing (such as shouting at people training), ending the requirements of exercise and uniforms.

The other area of forced labor that should be voted down is in prisons. If people want to work in prison voluntarily that's fine, but my own vote is that they should not be forced to work.

Ending discrimination based on age

Since age is not a clear guide for all people, laws based on age should be voted down in my opinion. Many times a group of people lose their rights, because they are not powerful enough to defend them, and this is clearly the case for young people. There is a mistaken belief that young people cannot decide for themselves, but yet, I think there is a lot of evidence against this belief. For example, even a baby can express a like or dislike for some particular food, song, movie, a hug, etc. Young people should not be denied the right to make decisions that concern their own bodies, to vote, and to consensually work- this was the big problem with the laws that allowed slavery and that denied women their right to vote, own property, to get an education, to work, etc. The view that humans reach adulthood around age 16-18 doesn't align with the physical reality of when young people reach the adult stage (puberty) which actually occurs around age 10-12. But just because a human is young or old (or excluded for that matter) does not permit the denial of their basic human rights and the requirement of consent, or excuse the use of molestation or violence against them. For example, I am shocked that, while spanking of a nonviolent child is clearly a violation of the assault law, spanking and belting of nonviolent children is very common- in fact I was even spanked as a child. I would ask your parents if they were spanked- I was surprised to find that as a child my Mom was terribly belted by her Dad, simply for telling a dinner guest "don't pick at our turkey". Even in many developed nations it is an outrage and illegal to show a young human a picture of the nude human anatomy, but acceptable to assault them with a belt or by spanking.

Total freedom of all information-no jail or fine for any info owned- the myth of "privacy"

Here's another simple social idea that seems inevitable as we move into the future, but the major newspapers, television shows, etc. nobody, will talk about this obvious "elephant in the room", "emperor wears no clothes" kind of truth: the idea of a society of absolute and total freedom of all information and the consequences of that kind of system.

Clearly with the flying nano-cameras and neuron readers and writers the ancient idea of "privacy" is totally a myth, except maybe for the privacy of neuron owners from the excluded. With the advent of remote control microscopic cameras, microphones and particle devices, which may have occurred as early as the 1300s, the concept of privacy became obsolete (especially for poor people). The wealthy already see inside our houses, so stopping the free flow of information can only limit the poor, and those who don't get to see. The myth of privacy, only serves to protect the monopoly those wealthy people who routinely see inside houses and heads have held for many centuries- and to protect those who murdered, assaulted, molested, stole and/or lied. The myth in many excluded people's minds that there is "privacy" is critical in order to get excluded people to do violence, theft, and inappropriate sex acts. If the excluded people knew the truth, that many millions see them, then, like so many D2B consumers, they never would follow remote suggestions to do violence, theft, or sexually inappropriate activity. Privacy is exactly the same as secrecy, and secrecy is wrong. Many violent crimes have only been seen by a small minority of people because of the widespread belief and support for secrecy and privacy. The much safer and smarter view, in terms of survival, I think, is the "leave no stone unturned" view- the view that nothing should be secret, and that secrecy is dishonest and evil. There is an obvious truth if owning or making public images of a crime is illegal because it violates a privacy, national security, pornography, or obscenity law, and that is that we cannot possibly stop a crime or catch a person who did a crime that we don't see, and that can only help those who did the crimes to stay unseen and unpunished to commit more crimes. Certainly when a person has done a violent crime, any "privacy" relating to images revealing the violent crime should be lost, and to keep such images secret is to be an accessory to violent crime. Opening up the free flow of information, in particular videos, is to close the space between the excluded public and the D2B owners and consumers.

One important reason to support total freedom of all information is because information is the only way to determine what the truth is. It's frightening how the major media and D2B consumers all lie about D2BW, Frank Fiorini, Thane Cesar, 9/11, 7/7, etc. You can see a potential even more frightening future where the D2B owners use their tremendous advantage to do more 9/11 kind of violence and massive lies, without ever been seen. For example, imagine the 9/11 crime without the videos of the three controlled demolitions; it would be much harder to prove controlled demolition. The same is true for the JFK murder without the Zapruder and Mary Moorman images, and for the RFK murder without the Noguchi autopsy; it would be much harder to know the actual truth. As if the block on the public seeing all the camera and thought images is not bad enough, D2B owners and consumers may even produce fake photos, for example in the case of the JFK murder and cover-up; the autopsy photos of JFK and the famous "Life" magazine photos of Oswald were altered. Beyond the probability of D2B owners and consumers doing more murders and cover-ups like 9/11, there is the very real possibility that people telling the truth about D2B, 9/11, the JFK murder, etc. might be jailed or hospitalized with made up crimes or psychiatric disorders by using D2B addicts to fabricate evidence and lie. Many poor D2B addicts can be easily made to lie in court in exchange for money or more D2B "services" like voyeur D2B videos, and free sex with beautiful women. I see this everyday with the D2B consumers that "shill" for money every 5 minutes. Probably the easiest way to remove those telling the truth about D2B, 9/11 and other lies is simply to galvanize them (remotely murder them). Those remote particle crimes must be very difficult to stop and to solve, and so total freedom of all information is very important for this reason alone. In addition, people telling the truth about D2B, 9/11, etc. and popular people that support full democracy ("political enemies", etc.) can easily be nonviolently removed from society by wealthy D2B owners and consumers using their tremendous D2B advantage. Hospitalizing or jailing people for made up non-violent psychiatric "disorders" or crimes is probably the easiest method to remove people from society, because you don't need any physical evidence like a corpse, bruise, or video, you only need a few D2B addicts to lie in court. Because of the lack of information, many times, even simply accusing a "political enemy" of some made up crime is enough, because the excluded people have no way of seeing the actual truth. Probably the easiest way to nonviolently remove people telling the truth from society is for:

5) A **drug crime**: D2B consumers in police simply claim they found illegal recreational drugs on the political enemy.

4) **Plotting violent crime**: the honest/political enemy is accused of plotting to do violence. No video or thought-images are needed; probably even a few fabricated text emails or photos of planted explosive materials can be produced in court. This is the classic Bush-era "terrorist" charge. Nelson Mandela was subjected to charges like this.

3) A **national security crime**: the honest/political enemy is accused of violating national security. A recent example of this is the arrest of "WikiLeaks" leaker Bradley Manning. How could the D2B not see any potential leaker when they have a massive system of nanocams and D2BW? Then since this is a nonviolent "crime", why can't people who leak just be simply "let go" or removed from classified areas without being jailed? Many times, the leak is legal and ethical because it reveals a crime or lie.

2) A **sex crime**: one or more (many times young) D2B consumers lie in court about being touched sexually. No physical evidence, like a video of the crime, is needed. Alternatively, child pornography can be (electronically) planted on a political enemy or produced in court. Even having some D2B consumers produce fabricated text emails or text chat is enough to convict a truth-telling enemy of *attempting* to do a sex crime, for example with a person under the age of 18. Even the accusation of a sex crime is enough to ruin an enemy's popularity and career. This kind of phenomenon is often called a "media assassination", because the person accused many times can never recover their reputation even when the claim is totally false.

1) A **psychiatric disorder**: one or more D2B consumers in police simply transport the political enemy to a hospital where they are hospitalized for life. Most people will not defend somebody accused with a psychiatric disorder.

Without the public having access to the images captured from the many tiny cameras on streets and in houses, and without the public having access to all the thought images and sounds captured, you can see just how easily D2B owners, consumers, and the major media can trick the excluded public with lies. But with all those images reaching the public, the opposite is true; tricking the public with lies and fabricated evidence is much more difficult.

Decriminalize recreational drugs

Simply put, we don't jail people for being overweight, for drinking too much alcohol, that smoke tobacco, that don't exercise, or for living unhealthy lifestyles, and we shouldn't jail people that choose to use recreational drugs.

If we do jail people for owning or using recreational drugs, let it only be for a few days, until they become sober and have another chance at sobriety- not for years as the current punishments require. People who sell ('traffic") drugs can even get a death penalty in some nations –it's absolutely absurd. Guns cause far more damage, and then to *other* people, and we don't jail people who sell guns.

There are many arguments for decriminalizing recreational drugs. This is not to say that I think recreational drug use is a good activity; getting addicted to recreational drugs, like smoking tobacco, and drinking too much alcohol, is definitely a bad idea. But violent crime is the big evil, not nonviolent crime, in particular where the so-called crime is people who are only hurting themselves. Billions of our tax dollars are spent on arresting, jailing, feeding, and clothing millions of self-sufficient, non-violent, otherwise lawful people. That money could be used to expose and stop violent crimes and theft.

Decriminalizing recreational drugs, takes the "caviar" aspect away from recreational drugs- the feeling that expensive products must be very desirable, and it takes the million dollar profits out of a violent and illegal market- just like ending the prohibition on alcohol did. It's idiocy to support a system where poor young nonviolent kids spend hundreds of dollars, and subject themselves to dangerous areas and people, in the search for illegal recreational drugs. The war on drugs should be a nonviolent, non-jailing war that uses facts, videos, and other information to get the truth (in particular about RNRAW and D2BW) to the public so they can make good decisions about what to put into their own body.

I'm glad to say that I do not use recreational drugs, although I did when I was younger, and my advice to people is not to get addicted to recreational drugs, tobacco, or alcohol. I admit that challenging a person's mind or getting a different perspective on the universe by using some kind of drug, like a psychedelic mushroom or marijuana, might have some beneficial effect, and there are probably some illegal drugs that might have some health benefits. Recreational drugs are many times used to cure boredom. But an addiction to tobacco, alcohol, and many drugs is like a terrible handicap and a ball and chain on a human that may last for many years, and that may be very difficult to stop. For example, I hate that I smoked tobacco for years- that was so stupid, and probably one reason why I didn't get a lot of dates. I think a lot of D2B excluded progressive-minded young people (which I was) are secretly and unconsensually remotely neuron written on with "ads" to start using alcohol, tobacco, and drugs at a young age – because it's a form of exterminating and making extinct progressives by the many conservatives that control D2B, but also because it produces more income for the companies that make and sell those products. Instead of trying to get a kiss, the excluded young people end up sucking on a bottle or cigarette. Partying with alcohol and drugs is a terrible and stupid tradition, but like the traditions of the religions, which are clearly based on lies and are also unhealthy and stupid, millions of people still believe in them and follow them, no matter how unpleasant and self-hurting they are. A drunk person is an opportunity for the remote neuron writers to make a person, especially a young sexually frustrated excluded person, do something stupid using remote suggestions, in particular violence and/or something sexually inappropriate. I think many people turn to alcohol, tobacco, and drugs not only because that is the wishes of the conservative remote neuron writers (and of those companies that sell alcohol, tobacco, and drugs), but because it's an alternative to physical pleasure, which in this time is callously forbidden. I think that people use alcohol and recreational drugs, not only because they are excluded and can't defend against remotely neuron written suggestions, but to try to escape from the terrible reality of our situation – where murderers and bullies run the show and the decent, smart, and honest are being trampled on by them – basically – this constant theme of the Inquisition seeking to torture and exterminate the young Galileos and Galileas of today. Many excluded people (progressives and non-religious) are remotely funneled into alcohol, tobacco and drug addictions and away from reproduction by the violent liars that control remote neuron writing. All most young people really want is physical pleasure- but because talking about and even educating people about pleasure, dating, kissing, etc. is not allowed, that natural desire is diverted into violence, alcohol, partying, sports, and other more acceptable non-pleasure/non-sexual based activities. In addition, pursuing sober intellectual pleasure and science is shunned by the religious conservative majority as being "nerdy", but yet, embracing and exploring science and technology seems like the logical path of the future, to wealth, friendships, and intellectual stimulation – really the exact opposite of religions (except for maybe friendships). So it's terrible that young people are not finding a large and solid tradition based not on religions, violence, sports, and partying with alcohol and drugs, but based on science, stopping, exposing, and punishing violence, showing and telling the stories of history, science, and the future, using their talents to express their views, celebrating physical pleasure, physical fitness, and other natural, sober-minded, and honest pursuits.

Decriminalize consensual adult prostitution

If we could see the direct-to-brain community, I am sure we would see a massive market of physical pleasure for money- perhaps the biggest money making market of D2B (or perhaps the buying of remote writes, or reads themselves gain the telecoms the most money). They apparently already have it down to a science with people getting "kiss", "hand", "sucks", "coatings", "holes", etc. I can only report from what I hear as an excluded. But beyond the fact that the D2B owners and consumers have already accepted a free and open market of pleasure for money, are all the moral arguments for not punishing people who do pleasure for money.

It's absurd to punish people who are involved with pleasure for money. First, these are nonviolent and consensual activities. One argument constantly echoed is that there is not consent. But agreeing to provide physical pleasure for money is exactly as consensual as agreeing to bag groceries, drive a truck, or fight in a boxing ring for money.

Beyond this, in the next few decades two-leg walking robots are going to be doing all manual labor (cleaning, shopping, driving, etc.). Humans doing physical work is going to become more and more obsolete. But even after many centuries, there will still be one job that humans are still hired to do – only one manual labor task will remain- not cooking, driving, or cleaning – but prostitution – providing physical pleasure. Not that the robots will not be very good at providing physical pleasure that looks and feels exactly the same if not better than a human, but certainly many humans will prefer to touch other actual humans for physical pleasure. So isn't that ironic that the only employment that will last long into the future is currently illegal?

I think that many educated people do not realize how harsh the society we live in currently is. For example, people (even children) can fight each other for free and for money (as sport), but even adults *asking* each other to pay or receive money for pleasure can result in being jailed. We live in a bizarre and shockingly frigid anti-pleasure group and tradition. We are all the product of sex- nobody can deny that – but yet many people are so violently antisexual. It makes no sense to curse the source of our life and physical pleasure. For some reason religions have always taken an extremely anti-pleasure based view. While violence is ok and manly, physical pleasure and even talking about physical pleasure is seen as a weakness. It's tough to know why this mistaken belief that physical pleasure is somehow wrong or a weakness occurred. Maybe population control was an important concern at the time. It may be that many people seek to stop others from enjoying pleasure out of jealousy. There is a popular view that sex and most forms of physical pleasure are extremely serious and life altering events that should occur secretly and only between two people that mate for life. But a more accurate view is that wanting physical pleasure is not a weakness, but is a natural and normal desire, that there is nothing wrong with the nude human body in public, and that most forms of physical pleasure (like sleeping together, fondling, kissing, or "handing") cannot result in pregnancy and so are not nearly as dramatically serious as many people make them out to be. So I think that in comparison to violence, people have come down overly harshly on no-chance-of-pregnancy nonviolent forms of consensual physical pleasure.

In the time we live in, people can be jailed even for asking to pay to touch a genital or to be masturbated ("solicitation"), or for even owning pictures of nude humans ("pornography"). For years people could be jailed for "seduction" (having sex under the promise of marriage). In many ultra-religious nations people who are suspected of homosexuality are routinely executed- it's bizarre. Violence is ok, but even talking about gentle touching is a big evil in their minds. But what we are seeing is a gradual enlightenment taking place on the Earth: the old laws that jailed people for homosexuality, adultery, seduction, etc. are falling because the public is becoming more logical and educated in its views of sex.

Many of the people in prostitution are poor people, and they could be using money from prostitution to pay for food and for an apartment. So those who are so vocal in opposition to decriminalizing prostitution, perhaps without knowing it, are actually helping to starve and make homeless many perfectly fine people, who otherwise might have had a job and would have survived.

Another point is that it seems illogical that making pornography, where people are paid to have sex, is legal, but prostitution is not legal. Clearly, there is almost no difference between if an individual pays for sex for themselves, or if a filmmaker pays for other people to have sex.

People buy stories in the media all the time to try to link the pleasure market with children, forced labor, and violence- which may happen, just like any market, but we don't hear about forced labor in other markets. There is also never any mention of the beneficial effects of legal prostitution: that otherwise unemployed homeless people are working and able to buy food and rent a room, that there is less violence because aggressive males have less sexual frustration, and that without the secrecy, people can more openly monitor and stop crimes and the spread of sexually transmitted diseases.

Just like advocating decriminalizing drugs, just because a person advocates decriminalizing prostitution, doesn't mean that they are involved in it.

In the time we live in, very few people openly advocate decriminalizing adult prostitution, because the stigma of being labeled a "pervert", "whore", or "pedophile" is so great, and because they don't want to appear to be a person who is involved with prostitution even if they aren't involved in prostitution. In addition I think that many wealthy D2B owners and D2B consumers may prefer keeping adult prostitution illegal because it may lessen reproduction among poor and D2B excluded people. Beyond that, the telecoms ("Madam Bell") probably want to maintain their secret monopoly over prostitution under D2BW; ending the prohibition on prostitution would obviously result in a loss of profit for them, as both sellers and buyers would have an alternative system. There are very few heroes to point to who stand against the bizarre anti-prostitution fervor. Many nations have already stopped jailing people involved in consensual adult prostitution (Wikipedia has a map assembled[85]). The National Organization for Women passed a resolution calling for decriminalizing prostitution in 1973[86], and in 1999 the United Nations CEDAW committee called for decriminalizing prostitution in China.[87]

---

[85] http://en.wikipedia.org/wiki/Prostitution

[86] Weitzer, R.J. Legalizing Prostitution: From Illicit Vice to Lawful Business. New York University Press, 2011.
http://books.google.com/books?id=cjJKlRQsEv0C&pg=PA249

[87] Pitcher, J., M. O'Neill, and T. Sanders. Prostitution: Sex Work, Policy and Politics. SAGE Publications, 2009.
http://books.google.com/books?id=_L-UkxjWVoAC&pg=PA101

Antipleasure ferver

One mystery about sexuality is how the vast majority of people alive are the product of two people who had sex, but yet a strong majority view exists that views nudity, the nude human anatomy, and public images of sex as being highly offensive; generally sex and sexuality are viewed as a bad thing. The obvious irony and paradox is: How do people so openly outspoken against pleasure and sex exist? If they were truly anti-sexual, they would not have sex, and not reproduce and therefore would quickly go extinct, but yet, here they are in large numbers. If they were so anti-sexual would they not be very frigid and cold and very difficult to cuddle up with, and therefore very difficult to have sex with, and wouldn't that always result in extinction of the anti-sexual person? So it's mysterious. It may even imply, that some aspect of why sex continues is deeply linked to dishonesty, trickery, and hypocrisy- that is, sexual success is linked to people who publicly support one view, but secretly practice the opposite view. This truth implies, for example, that to be a successful reproducer, you must pretend not only that you are not sexual, and not interested in sex, but that you are highly offended by all things sexual, like pornography, and must vigorously label others "perverts" in order to distance such labels from being attached to you. But then, in a passionate moment, suddenly, hypocritically you must throw away and betray all your antisexual views to make passionate love to some member of the opposite gender (kind of like Majors Burns and Houlihan in "M*A*S*H"). With all due respect to religious people and D2B consumers, of whom there are many honest people held hostage to tradition, religions have been breeding true for "liars" for

centuries and the D2B secret and lies have amplified
the success of the dishonest.

One clear truth is that there is an extremely
unnatural selective advantage for reproduction by
those who control and receive direct-to-brain
windows. Over the last few centuries, those who
received D2B must have been able to reproduce far
more easily than those who did not receive D2B.
Like wealth, being a D2B consumer plays a large
part in determining if a human will reproduce. The
problem is that progressive, non-religious, and
overly honest people are being systematically
excluded, and so perhaps that is one reason this
dishonest public antisexual fervor (but private sexual
fervor) is so popular: because dishonesty is a
requirement for receiving D2BW and receiving
D2BW is linked to success in reproduction.

Just like nonviolent atheists, agnostics, and
scientists were brutally punished in the Inquisition,
and other witch trials, and Jewish people in Nazi
Germany, so now in our time the "sex offender" is
the new heretic and witch. Shockingly, there is no
"violent offender" hysteria. For example there is a
"sex offender" list, but no "violent offender" list. Sex
offenders on this "black list" are not separated into
"violent" and "nonviolent", and they are all put
together on one list whether they have one or one
hundred sex offenses.

Remote neuron writing is currently being used to a
large extent with the goal of making sex offenders of
excluded. Using remote neuron writing to make a
person "bite" on an inappropriate sexual suggestion
is one of the easiest ways to lower the value of a
D2B excluded, and to put them on the path to
extinction. Excluded are constantly bombarded with
thousands of suggestions to inappropriately touch a
person under the age of 18, or to go outside in the
nude, etc. But the neuron writers are never viewed

as sex offenders- all blame is placed on the excluded human that bit on the suggestion.

One focus of the antisexuals of this time is on child pornography. Making child pornography illegal only protects those who abuse children. I think that making child porno illegal is also popular among the wealthy neuron owners because they want a monopoly on all information, and they are nervous about the public seeing images of what their eyes have seen, and the images on their thought-screens which may show their involvement in violent particle crimes. One interesting aspect is how AT&T and the telecoms always mysteriously escape all blame for crimes committed on their wires and with their equipment. For example, the government and public never blame AT&T for illegally transmitting child porn. Are we to believe that AT&T doesn't know what is on their wires and wireless network? Even if they don't (which is very doubtful given RNRAW), are they not partially responsible as accessories to the child porn crimes?

When we excluded see the news of young kids being killed in sex crimes (Chelsea King, Adam Walsh, Megan Kanka, Jon Benet, etc.) should we not wonder, as crazy as it sounds, "how could a child be murdered when people have been seeing and hearing thoughts with dust-sized cameras and particle devices for centuries?"? It can only be that the owners of remote neuron reading and writing are allowing these murders to occur. It may be that the view of many neuron owners and consumers is that sacrificing one poor child is a small price to pay, for the greater good of using the anger such a murder creates to pass new laws restricting the freeflow of information, and which create new options to jail or intimidate their enemies with made-up, hard to prove or disprove, career ruining, sex crimes. Much of remote neuron violent crimes are about causing a "wave of indignation" among the excluded public. Imagine if the public ever does get to see D2BW, how many people that were accused of made up sex crimes will be shown to be completely innocent, and how pissed they will be at all those people who actually caused all the crimes remotely using particle devices. One thing is clear, that the money people pay for sex and violence-filled videos fuels AT&T and the neuron. D2B consumers must pay millions of dollars to the telecom companies to see videos of the murder of those kids, their eyes, their thought-screen, the thought-screen of the murderers, their thought-audio, the thought-audio of the murderers, videos from different cameras that captured the murder from different angles, the sounds of the murder, the profiles of those who did the remote neuron writing, etc. – we can only guess- but I don't think we have to guess much.

Making child porno illegal is helping to protect many people, some of whom are probably D2B consumers and perhaps even D2B owners- because nobody in the public, in particular excluded people, can see their crimes.

With child pornography, like images of violent crimes, we need to remember that capturing or seeing evidence of a crime should never be considered a crime, only those who do a crime- in particular, a violent crime, should be punished. Protecting evidence of crimes, in particular violent crimes, is of extreme importance for people to know who really did the violent crimes. In addition, making sure that all the images can be easily seen by the public, and not just by employees of governments is very important. Thousands of important videos of crimes have been confiscated, kept secret, and destroyed by people in government, the Zapruder film, the Scott Enyart film, and the 9/11 videos, are just some famous examples.

It seems likely that many wealthy people want to fuel the public obsession with sex "crimes", in order to remove the more logical focus on violence. With the focus on sex crimes, the possibility becomes less and less of ever seeing images of the people that did the controlled demolitions of 9/11, and a million neuron murderers that currently live unseen and on the loose.

Trying to make child porn illegal is a disaster. For example, can you imagine that all the ancient vases showing child pornography might be illegal? We have the label "paedophile" but not "paedokrust" (a person who is violent to children), which shows the misplaced priorities. And what about child "krustography" or "violentography"? Is the next step to make capturing, viewing, and sharing images and evidence of violent crimes against humans under the age of 18 illegal? A perfect example of why nobody should be jailed for capturing or showing the public videos of a crime against a young person is the recent Hillary Adams video[88], which she posted to YouTube. The video shows her when she was 16 years old being beaten by her father who is a judge that has ruled on child abuse cases. There is a clear analogy between this video and videos of sexual crimes against young people. If people can be jailed and be on life-long offender lists for videos of crimes against young people, the public might never know that a crime was committed, and I think it's important for the public to know the truth. So I think that videos of crimes, even sexual crimes, against people of any age should not be prohibited.

Clearly humans start masturbating around the time of puberty; eleven or twelve years old. I was eleven years old was when I started masturbating – how old were you when you started masturbating? Masturbating is healthy and natural. What is unusual is if a person does *not* masturbate. For example, the practice of celibacy, common, in many religions, to me, seems biologically unnatural, unpleasant, and

---

[88] http://abcnews.go.com/US/texas-judge-caught-beating-daughter-tape/story?id=14867233#.UMDmlnewXTo

unnecessary. The vast majority of humans have masturbated, but yet most people do not publicly acknowledge that they have ever masturbated. Health professionals should be recommending that post pubescent people masturbate or have sex (with careful attention to pregnancy and disease) regularly at least once a week, (for example to masturbate or have sex every other day at some regular time), but they are mostly silent about masturbation, orgasm, and sex. Most people are not even told, for example, the theory that you should try to postpone your orgasm as long as possible in the last few minutes, to savor and have more control over the actual moment of orgasm. To deny young people the physical pleasure of kissing, and mutual masturbation is a form of neglect and child abuse because they are denied their right to consensual physical pleasure with similar-aged peers. Many people get a pet to substitute for this severe restriction, but it seems more natural to allow young people to kiss and fondle each other in a non-pregnant making way. The remote neuron has great pleasure in making young horny people do embarrassing things in desperation for physical pleasure- for example making them masturbate during class, or getting them to try to asphyxiate themselves during masturbation- which is a shocking and terrible thing that some young people do in lonely and isolated excluded desperation with unseen inhuman remote neuron writers constantly writing such suggestions for D2B consumers hungry for shocking videos. I am so sorry to say that many D2B owners and consumers must pay a lot to the telecoms to see videos of these poor young sexually frustrated excluded kids biting on terrible and radical remotely neuron written sexual suggestions. It's brutal, but you can't blame the D2B consumers, you need to blame those writing the suggestions and those enforcing total and absolute celibacy on minors- the very humans that probably need love and caressing the most.

It's scary even to talk about the anti-sexual fervor, because, like so many witch hunts of the past, any people who stand up against the injustice and idiocy to defend those labeled as witches are quickly labeled as being witches themselves, and publicly standing up for children's right to consensual physical pleasure with other similar-aged kids, and for protecting and making public images of crimes against children, is no exception. Those very few who are brave and disgusted enough to say anything are promptly and inaccurately labeled perverts and paedophiles, while the non-stop remote neuron, massive, billion dollar, secret, crime video-production, telecom industry continues production without missing a step.

I sometimes hear labels like "whore" or "slut", and it's evidence of a massive anti-women equality group, because sex is not just the responsibility of the woman, "it takes two to tango", but also, because a human should enjoy sex and orgasm once a day, and at least once a week- there is nothing wrong with having regular sex. A monogamous person may have just as much sex as a person (female or male) that has sex with a different partner each day, so there is no physical difference. But also with labels like "whore" and "slut", as a male, it's annoying, because, women don't need to be chastised and made colder in this ice age we live in- they need to be warmed up, turned on, complimented, celebrated (not celibated), encouraged to enjoy their bodies, and pursue pleasure soberly and intelligently. The "whore" people are usually the same people that constantly call everybody "gay", "fag" and "dyke". I'm so looking forward to the day when people realize that homosexuality is of no concern and is completely normal and natural. Violence is the big evil and should be the big taboo. Then, to those people, absolutely everybody is gay- it's so annoying- it's like homosexuality is their primary focus in life. I often say when hiring- "anybody but the violent, rude, and the anti-gay, for the love of the work environment." Defending gay people is just like defending people accused of being a heretic or witch- because you are promptly labeled gay by the idiotic antigay. In truth all people are probably naturally bisexual to a certain degree- there are only three kinds of people- those who admit to having masturbated to same gender touching, liars, or those who have had their minds chained with fear. Constantly labeling the excluded celibates "gay" and "pervert" has been a patented method used to stop the breeding of the honest for centuries.

Much of the anti-pornography, anti-sexual ferver is clearly anti-education, because how can you teach young people about anatomy and about the 600 million years of animal evolution through sexual reproduction when it is illegal to show young people images that contain a penis or a vagina? Do we want the next generation of doctors and other professionals knowing next to nothing about vertebrate anatomy?

There is a startling truth that most of the public is not told, and that is that we are descended from protists (single-celled organisms) that probably were very much like a sperm and ovum. Our sexual organs are the most primitive parts of our body. Before there was a brain, muscles, or intestine, there was a gonad. It's amazing to realize that all the other organs are, in some sense, "accessories". The entire digestive, nervous, muscular, skeletal, circulatory, respiratory, and endocrine systems are later developments; products of 600 million years of the elaborate mating dance, and evolved just because they are effective at bringing together sperm and ovum, those cells so like our primitive protist ancestor.

Consent-only health care (ending torture and unconsensual experimentation in the psychiatric hospitals)

It seems clear that the psychiatric system will eventually have to end all unconsensual electrocuting, surgery, drugging, and bodily restraints. The psychiatric system is a scary phenomenon, in particular because a person can just be picked up off the street and held indefinitely without any trial or crime being committed. Many people call the psychiatric system a "Siberia" after the famous destination of many political prisoners under Stalin. The first murdering by poison gas done by the Nazis was in psychiatric hospitals on "patients" (many probably pro-democratic, homosexual, and Jewish people) as a "hygienic measure". Holding people who haven't committed any crime or have only committed a misdemeanor, in a building against their will for an indefinite period of time is a violation of the Habeas corpus act - and that was progressive in the 1200s. Injecting people with drugs (experimental or otherwise) without clear consent, and physically restraining people (in particular nonviolent people) with four point limb restraints, to a bed for extended periods of time, leaving them unable to move their four limbs freely, and with no choice but to urinate and defecate on themselves, violates the ancient constitutional amendment and basic human value that forbids "cruel and unusual punishment", in particular where the punishment far out-weighs whatever crime was committed.

Just like lowering the popularity of an excluded can be done, very easily, by neuron writing constant suggestions to do something sexually inappropriate, so can making excluded do unusual things that get them locked in a psych hospital, which dramatically lowers their popularity, and attaches the lifelong labels of "psycho", and "nutter", etc. to them- even when their "crazy" act was the result of remote neuron writing, was nonviolent, and only a misdemeanor (like a temporary public outburst, etc.). Many of the people targeted for this kind of remote abuse are the honest and educated people of the D2B excluded. You can see how the 9/11 controlled demolition and Kennedy killers constantly try to label those telling the truth about those crimes as "crazy", "nuts", "kooks", etc.; trying to link the telling of truth with a psychiatric disorder. This is also the case for those that are figuring out remote neuron reading and writing, and starting to tell people publicly and honestly that they think people can hear their thoughts, that they hear voices in their head, that somebody is remotely moving their muscles, etc. You can see clearly how the abstract theories of psychology, coupled with the stigma of being labeled with a psychiatric disorder are being used to suppress the truth about remote neuron reading and writing, and to protect many murderers, like those who planned and carried out the controlled demolitions of 9/11, the Kennedy killings, and many other murders. They are people that have to use labels and theories of psychology, because the facts and physical evidence don't work in their favor.

Many of the claims based on psychology are very abstract and have little or no basis in physical science. Take for example, labels like "psychosis", "neurosis", and "manic depression". Unlike a broken bone, cancer, or virus like HIV, there is no bio-chemical diagnostic test to show that a person has psychosis, neurosis, or manic depression.

It's tough to state clearly what the basis of the "crazy" phenomenon and theory is. For example, one aspect is that a person who is labeled crazy has "inaccurate" beliefs. But we never hear the word "inaccurate" being used. The problem with locking up people with inaccurate beliefs is that, having inaccurate beliefs is not only completely legal, as it should be, but historically, a majority of people have always had extremely inaccurate beliefs- for years people insisted that the Sun goes around the Earth, the main claims of all the religions, that Moses parted the Red Sea, that Jesus turned water into wine and brought people back to life, are all obviously inaccurate- but we don't jail religious people, restrain and drug them, and try to deprogram them with the more accurate stories of evolution, science and our possible future as a globular cluster. I think that I've shown quite clearly that the claims of a big bang expanding universe with background "radiation" are inaccurate, and that light is material and not an "electromagnetic" wave, and so even many currently popular claims of science are inaccurate and "crazy"- but physical restraints, injections and other brutal punishments are not the answer- showing the public good information is a far less destructive and far more effective answer for correcting inaccurate beliefs.

Another aspect is that a crazy person is thought to have unusual and unpredictable behavior. Some people are somewhat unpredictable, or have inconsistent behavior, but as long as they are not violent, that shouldn't be viewed as being reason to be hospitalized. Much of psychology is geared towards removing creativity and difference in society. For example there is an attention deficit disorder, but having an attention surplus is not a disorder yet. The labels and "disorders" of psychology favor the dull; being overly dull is not a disorder. Beyond that, many people who do unusual, or unpleasant things are the victims of remote neuron writing, and would not do unusual and unpleasant things if there were no evil idiots writing terrible suggestions on their neurons.

In light of knowing the secret of remote neuron reading and writing, and that much of thought is simply thought-images and thought-sounds, most of the theories of the famous people in psychology (Freud, Jung, etc.) about "id" and "conscious", etc. have to been seen as far removed from being accurate. No psychology textbook talks about thought-screen, thought-audio, etc. – but remote neuron reading and writing has been known by many of those authors for over two hundred years, so clearly, without talking about remote neuron reading and writing, much of the theories of psychology are mostly fraud or too abstract and outdated to be of any value.

Many times the person labeled with a psychiatric disorder is a nonviolent person to begin with. Ultimately, in my mind, violence is the big evil, not non-violent activity, and those who do violence should be jailed, not hospitalized. There can be consensual treatments offered to violent people; for example, I think one good video to show them might be about how excluded people can be remotely neuron written on to do violence for some D2B consumer who doesn't receive any of the blame.

How about the so-called cures offered for the "crazy": drugs, physical restraint, etc. Anyone that sees a psychologist is going to be prescribed drugs (which funds drug makers). Everybody that is taken to a psychiatric hospital gets drugs; as if a drug is going to magically teach a person about the history of science, about how people write to their neurons, etc. Then think of the bizarre system we have: people who use recreational drugs consensually are brutally jailed for years, but those who just say "no" and refuse psychiatric drugs in the hospitals can be injected with drugs against their will. Almost every psych hospital routinely uses four point restraint torture without consent. This is simply a "punishment", not a treatment, and is clearly designed to make a person less aggressive and more submissive – many times because they have to beg "please untie me" and "please loosen these straps", and "please let me use the bathroom", etc.

It's shocking that those who murder can avoid prison by pleading "insanity", and then can be released as "cured" from some abstract and unprovable "psychiatric disease". The "insanity" defense is ridiculous in my opinion. My vote is for people to be held accountable for their crimes no matter what their motive was. In addition people who work in a hospital should not be subjected to violent people. We need to clearly separate prison from hospital and violent from non-violent. Beyond that, all health care should be consent only.

Many times, remote neuron writing is used to make people do inappropriate things (yell in public, talk back to a person in police, go out in the nude, do sexually inappropriate things, etc.). It will be interesting to see how many of the people locked in psychiatric hospitals are there because they are D2B excluded and bit on some remotely neuron written suggestion. Probably the vast majority of people locked in psychiatric hospitals are D2B excluded- people who know absolutely nothing about remote neuron reading and writing. So in terms of treatments, wouldn't that be the first thing to tell them about? "You know it may be that some terrible violent criminals are able to remotely write to your neurons...and it's best if you realize that any voice you hear or image you see in your thoughts is not from God but is from these really terrible criminals of the 9/11 demolition Kennedy killing remote galvanizing kind...". But that's not being done.

One kind of funny, but dangerous truth about the harsh, fraud-filled, time we live in, is that a person can never simply be "wrong" or "inaccurate" on some claim- instead, in the eyes of many people, having a mistaken view is always apparently symptomatic of a systematic psychiatric disease that results in them being "crazy". Only rarely is a person ever viewed as just being wrong on a few theories.

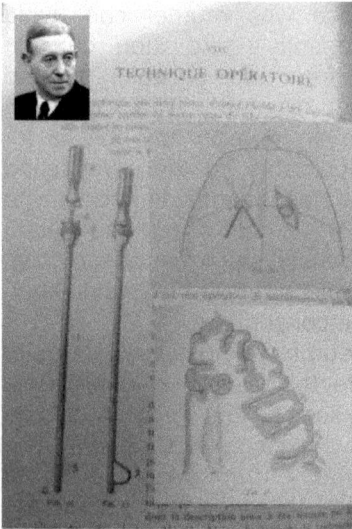

Figure 45. Nobel Prize winner Egaz Moniz and images from his paper detailing the first use of the Lobotomy.[89]

Beyond all this, the history of psychology is terrible. One bizarre story is how the person who first performed the "lobotomy" (perhaps unconsensually) on some people, Egaz Moniz (fig. 45), won a Nobel Prize for the "leucotomy" (lobotomy)[90] in 1949, at a

---

[89] E. Moniz, "Tentatives opératoires dans le traitement de certaines psychoses" (Tentative methods in the treatment of certain psychoses), Paris : Masson, 1936.
also in:
J Am Med Assoc. 1937;108(21):1828.
http://jama.ama-assn.org/cgi/content/summary/108/21/1828-g
[90] "Egas Moniz - Biography". Nobelprize.org. 29 Oct 2010
http://nobelprize.org/nobel_prizes/medicine/laureates/1949/moniz-bio.html

time when the truth about how psychiatric hospitals were used to euthanize innocent people in World War II, and the newly created "Nuremberg Code" advocating consent-only experimentation were still fresh in many people's minds. The lobotomy is a procedure that simply crudely removes parts of the brain. Even today, nonviolent, lawful, humans are subjected by law to being electrocuted with the radical and doubtful treatment of electroconvulsive therapy. For centuries people accused with abstract and unlikely "disorders" have been subjected to bizarre and brutal "treatments" like being restrained in a tub of water, having teeth pulled, etc. There are numerous books on this history, but no major psychology books or professionals want to educate the public about the shockingly brutal and violent past of psychology. Things are starting to change, however, as more victims are telling their stories, and more videos are reaching the public- for example there is now even a museum showing the history of psychiatric torture in Los Angeles. What we really need is a major motion picture or mini-series on national television that gently and inoffensively describes a lot of this truth and suppressed history to the excluded public.

Let the record reflect that in our time, labels of psychiatric mental disorders were used to suppress many great scientific and criminal truths, like the truth about the big lies of our time: remote neuron reading and writing, the "red shift" of the "expanding" universe, and the murders of JFK, RFK, 7/7 and 9/11. All those telling the truth are dismissed with abstract, meaningless, labels like "crackpot", "crank", "bonkers", "wacko", "nut", "kook", etc. Simply put, we live in a time where truth is called crazy and crazy is called truth. The murderers and their accessories have to draw on these abstract labels, because the facts and physical evidence don't support their inaccurate claims.

A "one-letter-equals-one-sound", one-stroke, easy-to-write, democratically-determined phonetic alphabet

We live in an era of massive language mispronunciation, corruption and misspellings, because we do not use a "one letter equals one sound only" alphabet. So as a result, there is a constant mystery, in particular, for young people just learning language, about how to pronounce any word. For example, should the letter "a" be pronounced like the "a" in "ape", or like the letter "a" in "apple"? Another result is that people are inaccurately pronouncing words like "Caesar" (Greek: "Καίσαρας") and "circle" (Greek: "κύκλος") because those words originally had a "k" sound, but were then changed to use the letter "c" which can have two sounds. When we make a spelling mistake many times people may look at us like we are stupid and uneducated, spelling correctly is many times viewed as a measure of intelligence, but isn't it more stupid to reject a one-letter equals one-sound alphabet that would, at once, end the vast majority of spelling mistakes, and provide more ease in spelling and reading for all people? A one-letter-equals-one-sound alphabet would solve all of these frustrating problems: the actual pronunciation of words can be more accurately stored over the years, kids can more easily learn language without wasting precious hours learning unique spellings, and questions of how to pronounce any word would be instantly removed. In addition, since all languages humans speak (Arabic, Chinese, English, French, German, Hindi, Italian, Japanese, Javanese,

Korean, Russian, Spanish, Vietnamese, etc.) all basically use the same exact sounds (the "a" sound in "ape", the "B" sound in "ball", for example), a single alphabet can be used for all human languages. In fact, this truth is a piece of evidence that all the 30 or so basic sounds of all human languages originated in Africa before the common ancestors of all Homo sapiens now living moved out of Africa into Eurasia (perhaps 100,000 years ago), and then on to Australia and America, because human communities separated by large distances have different words, but not different base sounds. A single international easy to use alphabet could simplify communication between people of different languages. The stories of every language could more easily be understood by all the people of Earth, because the task of learning the sound of each word would be removed. There is an International phonetic alphabet, but those letters are not single letters (for example "eye" is represented with two letters "ai" instead of simply "I"), and not single stroke letters. The fact that sound frequency can change the meaning of words in Chinese adds complications, but perhaps people can figure out a simple and logical system for that, like adding one of five slanted lines before or over a word. Here is a sample one-letter-equals-one-sound alphabet I made a few years ago:

ABDEFGhⱵJKLMNOPRSTUVWЧⱿACꓭLⱮJLMNⴱ
ꓭSⴱⵝⵐⴲ

I designed the letters to be writable with just one stroke. Just 40 sounds and their symbols cover most words of all human languages. To hear the sounds of each letter, see my web page tedhuntington.org/fonik.

Some words with their native spelling, and with a one-letter-equals-one-sound phonetic alphabet spelling are listed here:

Language	Native alphabet	Phonetic alphabet
English	cat	ꓘAT
English	teach	TƐC
English	hello	hɘLO
French	bonjour	BⵔNꓤUR
Chinese	你好	NƐ hⵔU
Russian	привет	PRƐVꓬꓭT
Arabic	مرحبا	MⵔRhⵔBⵔ
Persian	سلام	SⵔLⵔM
Hindi	नमस्ते	NⵔMꓭSTA
Italian	ciao	CⵔU
Vietnamese	chào	CⵔU
Japanese	こんにちは	ꓘONƐCƐWⵔ
Korean	안녕하세요	ⵔNꓬⵔNhⵔSAO
Hebrew	שלום	SⵔLOM
Greek	γειά σου	ꓬASU

It's tough to know how thought-images and sounds might replace text as D2BW becomes more widespread- an image can many times be a faster way to communicate information than with text- and certainly there is no need to create thousands of symbols to represent each of the infinite number of sounds possible. It seems likely that instant translation of all major languages can be written to the eyes (or ears) of those who are not denied D2BW, and probably soon walking robot assistants will translate and communicate directly with people of different languages for their human owners.

The other species, their thoughts, vegetarian alternatives

Of the many monstrous results of keeping D2B secret, is that all the thought images and sounds of the other species have been kept from the public for centuries. Which species can form thought-images is not even publicly stated yet. When a D2B consumer walks by any of the other species, a bird, dog, cat, horse, etc., probably they see the little thought screen and hear thought audio too- like a dog remembering a song, or the eye image of a bird looking at you. Seeing the thought images and hearing the sounds of the other species might probably lower the popularity of eating them by humans.

It's terrible that most people have not even been told such simple truths as: "leather is the skin of a cow", and "meat is muscle". I myself didn't make those two connections for years.

I think that some tastes are adapted to, because I remember clearly how the first time I ate a hamburger, after the first bite, which tasted like hay, I asked my Dad "This is what everyone eats?" in disbelief. I have been happily vegetarian for over 10 years and now after a massive rise in the popularity of the vegetarian diet, almost every meat product has a vegetarian equivalent at a similar cost. I think that if there is not a big difference between the meat and veggie food item, then of course, it's better to have the veggie food item because no animal is killed or enslaved for it.

An actual logical non-pseudo-science way to lose weight

There are so many myths and pseudo sciences surrounding weight loss, and many remote suggestions of images and smells of food, that I thought I would tell excluded people the system I used to lose 100 pounds. It's extremely simple: I only eat two meals a day spaced about 12 hours apart (6am and 6pm) with absolutely no other snacks except gum and any liquids I want to drink. I eventually even stopped the gum chewing habit by recognizing and sometimes "firing back" in my mind at each (presumably) remotely written impulse/reminder. One added benefit is that I am hungry in the morning and that is a strong motivation to get up and out of bed even if I'm still tired.

There is a lot that motivates me to get into good shape: to look attractive, to have a longer life, to attract a female partner, to increase the chance of getting affection (kissing, dates- already a very low probability for an excluded, in particular in this bizarre anti-pleasure age), to enjoy sex, (such that it exists in this ice age era), and to make a family.

When I eat, I eat whatever I want- but initially I actually counted bites. It's a simple equation: your body can process maybe 100 bites of food a day (50/meal), so if you eat under 100 bites of food you will lose a little weight, if you eat over 100 bites you will gain a little weight. I stopped counting bites, but a camera and computer could make it more convenient and "bite count" is good info to track along with weighing your body every day to know if you are gaining or losing weight. The key is to take small bites, and to savor the food. Since there is so little food, it tastes better when eating. Sometimes I imagine eating whatever I want to and it's amazing how similar to the real experience it is. In addition, I find that this greatly reduces my budget for food, so that I can actually spend more money on more exotic and interesting food items. Then I do a very minimum exercise every day of simply jogging 1 mile and sit-ups. On alternate days I also exercise my chest and side muscles. The key is stretching the muscles where the fat is. For males it's almost exclusively the abdomen and side muscles – why oh why are we not told this simple truth? For females probably repeatedly using the buttocks muscles is the most efficient method to lose the most weight from the area with all the fat. It's pointless to focus on muscles that don't have fat built up if the goal is simply to lose fat. You know that a particular exercise is effective when the muscles where the fat is become sore.

Reality of bipedal robots doing all manual labor

I talked about this earlier, but it is something that people have to understand and plan for. The idea of a "job economy" where humans work is falling to the past. Currently, humans are born, go to school, get a job, make a family, and then retire. But in the future, probably people will not have to get a job. It seems certain that in the future, clearly, robots will do almost all manual labor (cleaning, planting, harvesting, packaging, driving, shopping, etc.). How and when is not clear, but perhaps humans will democratically create some minimum standard of living, for example: all humans get a room, one meal a day, a free set of clothes, etc. If a person wants more, then they will have to find some way of getting money. You can sense how D2BW is already a massive money market (for those who are provided with it) that pays people even to simply say things or go places. Perhaps pleasure for money will be one of the few jobs still open to humans, but the vast majority of manual labor tasks will be done by low cost bipedal (humanoid) robots. On the plus side, many of the repetitive mindless assembly line tasks now being done by humans will be done by machines. Jobs like that are much better suited for machines – it's cruel to subject humans to that kind of work by keeping robots a secret. Instead of having regular jobs, people may get money from the government just to live on, in addition to money from wealthy people for their votes or because they have similar views and want to promote those views, etc.

Other popular mistaken theories and beliefs

Throughout history there have been many mistaken theories and beliefs, and my desire is not to put-down, or ridicule those who believe inaccurate or alternative theories, but to try and reach people with what I think are the more accurate theories. One classic theory was that Earth was the center of the universe, which of course, has fallen to the belief that Earth is only a small planet going around a Sun which is one of the many stars in the Milky Way Galaxy. The theory that all of space is filled with an "aether" fell out of favor, and I think the big bang expanding universe theory will also eventually fall out of favor with the public- in particular when they see that many of these popular claims are all part of the "big neuron lie"- many people lying because they are held hostage by the D2BW owners. Once you realize that many of these explanations from those in the media, in science, in crime solving, in religions- in every field- are part of that big neuron lie, then you realize that the actual truth of many events and the universe in general may be very different from the popular explanations told to the excluded public.

Here are some other popular mistaken theories and beliefs:

Mistaken Belief: That God or Gods exist

Believe in a God or Gods if you want to- all people must be free to think and believe what they want to freely. Without trying to upset or offend anybody, there are good arguments that suggest that no God or Gods actually exist:

- For centuries there was only polytheism (belief in many Gods), so you can see, historically, the origin of the "God" theory (the theory that a group of Gods control the universe), created by humans sometime less than 100,000 years ago. Monotheism, the theory that only one God exists also developed recently, around 3,350 years ago (1350 BC) starting with the Egyptian pharaoh Akhenaton.[91] Judeism is no more than 3000 years old, Christianity only 2000 years old, and Islam only 1400 years old.

- There have been so many Gods throughout history, and each religion has their own God or Gods, how could only one be the correct God? As one smart statement says- we are all atheists of some God; atheists have just rejected belief in one more God. I often say that belief in atheism is more conservative than Christianity and Islam because atheism is older than both those "new age" religions.

- We should interpret the universe in terms of matter and space. We should only believe in those things that can be observed with our senses.

This is not to say that there is nothing that does not defy logic. For example, that there is no beginning or

---

[91] "Akhenaton." The Columbia Electronic Encyclopedia, Sixth Edition. Columbia University Press., 2012. Answers.com 27 Oct. 2012. http://www.answers.com/topic/akhenaton

end in time or space in the Universe, or that galaxies and living objects exist at all. And this is not to say that I do not have the deepest respect for the Universe, which we are a tiny part of, for all matter and space, and all that we have learned about the Universe. I feel a strong and natural need to be true to my beliefs and feelings, for example I refuse to lie about D2B, 9/11, and my scientific beliefs. In addition I am in awe of the Universe and all that science has shown us. What most people call "God" would probably most closely map to "the Universe" in my value system.

There is a clear and obvious link between the many Gods of the polytheistic religions and the single God of the monotheistic religions. Polytheism is obviously the ancestor of monotheism. But this clear relationship is made less clear by referring to a single God simply as "God" without a quantifier like "a" or "some", because saying "God said this" instead of "a God said this" removes the possibility of their being more than one God, and seeks to hide the past history, and the original theory of Gods as a people who live in the clouds, and inside the Earth, etc., and that control nature.

One other thing that I find ridiculous is how many people promptly pray whenever tragedy occurs. We can't wait around for a God or Gods that don't exist to solve our problems; we have to solve our problems ourselves. Here we are reaching an age where the public is going to be communicating to each other with thought-images and thought-sounds, but there are still so many people that believe in these extremely ancient and inaccurate theories and traditions, like the ancient inaccurate theory that massless human-like Gods control everything in the universe.

Mistaken Belief: That a Heaven or Hell exists

Many poor people are terrified by the thought of a Hell, but it is obvious to me that the myth of Hell is a recent invention, and that no Hell exists in the Universe. If you say that something is "old as Hell", you are saying that something is actually very young relatively speaking, because Hell as a concept is no older than 1000 B.C.E. We can't close our eyes and hope evil goes away; if we really want to stop evil, we have to be brave enough to examine it. I encourage you to research this history too; anybody can see the truth of this simply from recorded history. The myth of Hell was adapted from the earlier myth of an "Underworld" ruled by the God "Hades" (and referred to as "Hades"), which was located underground in the Earth. Hades is the ancestor of Hell. Unlike Hell, Hades was not just for bad people, but was simply (and inaccurately) thought to be where people go when they die.[92] Judaism, which Christianity was initially a form of, had a similar concept called "Sheol" where dead people were thought to go after death,[93] and that is what the modern theory of Hell is descended from.

I recently learned that the infamously scary and supposedly unlucky and evil number of the "beast", "666" was actually, before Christianity, thought, by Greek people, at least, to be a "lucky" number, in particular a lucky dice roll of three sixes, a "triple

---

[92] "Hades", The Concise Oxford Companion to Classical Literature, Oxford University Press, 1993, 2003. *Answers.com* 16 Sep. 2012.
http://www.answers.com/topic/hades
[93] "Hades", *Encyclopædia Britannica, Encyclopædia Britannica Online.* Encyclopædia Britannica Inc., 2012. Web. 16 Sep. 2012
http://www.britannica.com/EBchecked/topic/251093/Hades

six".[94] Isn't that a funny truth? Oh a 666! It's my lucky day!

Beyond that, rejecting religious theories is no big evil. First strike violence against non-violent people is the big evil on Earth; nonviolent activity can only be a lesser evil. But rejecting false claims, religious or otherwise, is no evil at all, and is a great good.

The myth of a Heaven has a similar story. For centuries before Christianity and Islam, Heaven was the home of the Gods; mere mortals like humans were not destined to go to Heaven after death (except for a very few "heroes").[95] Only later did this myth change to make Heaven a place where "good" people go, and Hell a place where "bad" people go after they die. How can so many people accept so primitive and inaccurate a theory as being "divine" or sent by God, in particular, knowing that humans recently changed the theory and "renovated" Heaven?

---

[94] Aeschylus, The Agamemnon, Choephori, and Eumenides of Aeschyles, tr. into Engl. Verse, 1865, p4.
http://books.google.com/books?id=CIYCAAAAQAAJ&pg=PA4
[95] "heaven", *Encyclopædia Britannica. Encyclopædia Britannica Online,* Encyclopædia Britannica Inc., 2012. Web. 16 Sep. 2012
http://www.britannica.com/EBchecked/topic/258844/heaven.

Mistaken Belief: That a Devil exists

While there certainly are many bad humans living on Earth (first strike violent people, and those who constantly remotely molest and lie), there is no Devil, Demon, Angel, or God of evil that is the supervisor and controller of all evil as we humans define evil on Earth and in the Universe. Many people may not realize this, because the subject matter itself frightens many people, and many people that do examine it are wrongly and unfairly "demonized", but like the theory of a Hell, the theory of a Devil is a recent invention. The earliest mention of a Devil is of a "Satan" in the Hebrew Bible around 600 B.C.E., in the book of Job, and initially the Satan of the Hebrew Bible works under God's supervision, and is not the ruler of all things evil[96,97].

[96] Pagels, "The Origin of Satan", 1995, p39.
[97] "job new 2". The Columbia Electronic Encyclopedia, Sixth Edition. Columbia University Press., 2003. Answers.com.
http://www.answers.com/topic/job-new-2

Mistaken Belief: That Jesus was the son of God, or a part of God, or was supernatural

- Jesus really made no significant contributions to life on Earth, and in particular science. People who lived before Jesus gave the world pottery, the wheel, writing, the correct size of the Earth, etc. and humans after Jesus did much more, inventing the electric light, the automobile, the airplane, the camera- we have benefitted much more from their hard work than from anything Jesus, Muhammad, Abraham, Siddhartha Gautama, or Confucius did.
- It is ridiculous to idolize and view as relevant to today the teachings of a person that lived over a thousand years ago, in a time when there was no electricity, running water, Internet, airplanes, thought-recording, etc.
- It seems not very smart to view a human that lived 2000 years ago as being somehow radically different from the trillions of other humans that have ever lived- in particular when you think of the last 200,000 years of human reproduction and expansion.
- It's extremely Old Worldly to live a person's life devoted to a guy who lived thousands of years ago. Similarly, most people don't wear animal hides and use stone tools anymore.
- Many people have not really examined the history of Christianity. For example, if Jesus lived, he was definitely a Jewish person practicing Judaism. Few people recognize that Christianity is a form or "sect" of

Judaism. Christianity was initially Judaism and then made later adaptations.

- History is filled with the stories of violence done against innocent people in the name of Christianity. The Inquisition was used to inflict very violent punishments, like burning people alive. Most of the time the victims were nonviolent people, and often the brightest and most honest people of Earth (like Galileo and Giordano Bruno). In addition, violence and discrimination against people just because they have Jewish ancestry is a common brutal and illogical theme throughout the entire history of the Christian religion. Any Christian anti-Jewish racism is ironic and illogical, in particular because the founder of Christianity was a Jewish man.

- Jesus may not have even existed- other people that lived during the same time left writings like Pliny, Livy, and Strabo. If Jesus did live, he left no writings, and perhaps didn't even know how to read or write.

- Many of the claims of Christianity are obviously false, like the claim that Jesus brought people back to life, that Jesus visited people after his death, etc.

- There are great humans who made tremendous contributions to life of Earth, but it seems foolish to spend our life focused on a single person. How better it seems to me, to focus on the big picture- of evolution, the history of science, what we need to do to go to the other stars, and what our future might be like, but then also to focus on our own personal life and pleasures. We shouldn't torture ourselves doing activities that are not fun and interesting. Isn't time better spent trying to find a good mate, a good meal, and to pursue things that really interest us and that give us the greatest pleasure?

Mistaken Belief: That Muhammad was a profit of God

- As is the case for Jesus, Muhammad was just another one of the billions of humans.
- Almost all the same above arguments explaining why it doesn't seem smart to spend a person's life centered on a human who lived over a thousand years ago, apply in the case of Muhammad and Islam. Many other people have contributed much more to science and to making life easier for many humans than Muhammad did. It's absurd to worship and view a person's writings made in a time without electricity, airplanes, computers, etc., as relating to modern life.
- Having to bow to Mecca five times a day, like prayer, has no actual benefit, and is a tremendous waste of precious time.
- As is the case for Christianity, there is a lot of violence done under the name of Islam, and like Christianity, the violence done under Islam is often directed against the brightest minds, and against nonviolent people for trivial nonviolent "crimes" like blasphemy, adultery, etc.

Mistaken Belief: That many claims of Judaism, Buddhism, Hinduism, and other religions are accurate and useful

- Many claims of Judaism are obviously false, like that the Universe was created in seven days, every part of the story of Adam and Eve, the claim that Moses parted the Red Sea, that people must follow specific rituals to please the God, that sending thought-audio messages to God (prayer) will help to solve problems, etc.

- In terms of Buddhism, many claims are simply superstitious and almost certainly false, for example, that putting gold leaf on a statue of Buddha will bring good fortune. Like a "lucky four-leaf clover", there is no truth to those claims, and it doesn't help to participate in and perpetuate those inaccurate beliefs.

- The theory of reincarnation, one claim in Hinduism, seems very doubtful to me. I also doubt the theory of Karma, although there may be some truth to the idea that a living object may receive collective benefits or losses as a result of their individual actions- if ever humans on Earth choose to suddenly embrace logic.

I encourage people to learn about the details of evolution (for example in the books "Prehistoric Life"[98] and "The Ancestor's Tale"[99]), the history of science (for example in "Asimov's Biographical Encyclopedia of Science and Technology"[100]), and

---

[98] Palmer, et al, "Prehistoric Life", DK Publishing, 2009.

[99] Dawkins, "The Ancestor's Tale", Houghton Mifflin Harcourt, 2004.

[100] Asimov, "Asimov's Biographical Encyclopedia of Science and Technology ...", Doubleday, 1982.

basic history (like "Compact History of the World"[101]).
It seems obvious to me that science is a much more
honest, logical, accurate, and interesting philosophy
to embrace, follow, and participate in, than religion
is.

[101] Parker, "Compact History of the World", Barnes & Noble Books, 2002.

Mistaken Beliefs: Superstitions

Superstitions are terrible, and all of them are simply false. For example some classics are: a black cat crossing your path is bad luck, breaking a mirror causes seven years of bad luck, a severed rabbit foot is lucky, certain minerals have healing or magical powers, Friday the 13th is unlucky, opening an umbrella inside will bring bad luck, and that crossing your fingers will bring good luck. There is no logical basis for any of these superstitious claims, and we shouldn't perpetuate them.

Many other mistaken popular beliefs

There simply are a lot of people who create lies or mistaken theories, and a lot of people who believe them. That many humans are easily tricked into believing lies and mistaken theories has created many mistaken labels or nouns for objects that simply don't exist in the universe. Just to name a very few: Gods, Heaven, Devil, Hell, Witches, Soul, Goblins, Ghosts, Angels, Hades, Zombies, Santa, Fairies, Magic, Luck, etc. Many of these old beliefs were formed long ago in the past by people who had never flown in a plane, had no electricity, running water, or telephone, etc.; people with very primitive views compared to modern times. To a certain extent, some ideas were precursors to science- for example the ancient concept of "soul" was created before people modernized anatomy, identifying the heart, the brain, etc. But there are also many non-existent claimed objects that were the product of science, for example: an aether that fills all space, phlogiston, n-rays, and black holes. I don't believe in the existence of any object (or noun) for which there is no physical, observable evidence. In the absence of any physical evidence for an object, I think the smartest view is to presume that it does not exist in the universe, and that some people long in the past mistakenly thought it did, or simply created a pretend creature or place.

I'm a somewhat slow person. For example, my "ULSF" project has taken 8 years to complete. But, 700 years of waiting for RNRAW and D2BW to be made public is absolutely a shocking, ridiculous, and torturously slow delay. The same is true for the fact that 2000 years later many people are still talking about Jesus as being relevant to planet Earth now. It's time to update and modernize our society, and at a much faster pace!

Excluded people have been left uneducated, tricked, and lied to, and so many believe the supernatural religious claims and/or the many very unlikely popular scientific claims. But I think that the time is coming where many excluded people, are going to learn the extent of how they are being lied to, and get much better at detecting lies and determining who usually tells the truth and who usually lies. Many times those poor victims of all the lies and denial of service are the most willing to listen, while those who get D2BW generally try to avoid those who don't. Most of the D2BW consumers don't like to talk to excluded people because they already know that most of what the public is told about crimes, religions, and scientific theories are deliberate lies, and because they have already seen the thought-images and heard any thought-sounds that a person denied D2BW might want to share with them.

We can only imagine how many amazing, beautiful, and important truths and people have been ignored, or have never been seen or heard because of the popularity of the inaccurate theories of religions, psychology, fraudulent "science", because of the neuron lie, and because of the widespread belief in the myth of privacy.

Conclusions – What can we do?

To conclude this work, I want to say that I think that it's important to learn as much as we can about science, and the universe. It's an amazing set of truths that are being revealed about how life evolves around stars, how the most adapted life forms star clusters, and then moves around the galaxy like a giant fish in an ocean of space. I think it's important to understand the details of evolution, the history of science, and in particular the story of our future in order to help ourselves. It also helps to know about the possibility of secret technologies, like nano-cams and remote neuron writing, to have an accurate picture of what is happening around you, and in order to defend yourself against the abuse of such technology. Beyond our own understanding, it's of critical importance to show and educate the public about all of these truths. It's amazing and sounds unbelievable, but yet it seems very likely we will soon be able to see and share millions of thought-image and thought-sound recordings of many different species, including are own species, and then, sent directly to our eyes and ears. By knowing about our past and our future, we can start to plan for those inevitable events of the future, like when robots can shop, cook, and clean, when helicopters fly in air highways, when Moon and Mars cities contain thousands of big buildings, and when ships of ours can finally reach the other stars. We are constantly learning new truths, and exposing old mistaken beliefs and lies. Learning the truth about mistaken beliefs and lies helps us to stop wasting precious time on meaningless and useless activities, and to start finding more real, smart, and meaningful enjoyment in our lives. When we see our place in the timeline of history and our hopeful future as a star cluster, we must come to the logical conclusion that we should not waste time on mistaken beliefs and useless practices, or on tricking and lying to others, or leaving them in total ignorance, but instead, to start moving forward together into that exciting and interesting future.